Randolph H. S. Churchill

Men, Mines and Animals in South Africa

Randolph H. S. Churchill

Men, Mines and Animals in South Africa

ISBN/EAN: 9783744751223

Printed in Europe, USA, Canada, Australia, Japan

Cover: Foto ©berggeist007 / pixelio.de

More available books at **www.hansebooks.com**

THE R.M.S. "GRANTULLY CASTLE," 3489 TONS, IN DARTMOUTH HARBOUR.

MEN, MINES AND ANIMALS

IN

SOUTH AFRICA

BY

LORD RANDOLPH S. CHURCHILL, M.P.

NEW EDITION

LONDON
SAMPSON LOW, MARSTON & COMPANY
Limited
St. Dunstan's House
FETTER LANE, FLEET STREET, E.C.
1895

PREFACE.

At the request of the publishers, I have, against my own judgment, consented to revise the letters from South Africa which I wrote to *The Daily Graphic*, in 1891, with a view to their publication in the form of a book. The critics of literary and epistolary efforts, who daily inform the public through the columns of the Press, pronounced with tolerable unanimity, that these letters of mine were devoid of merit and unworthy of perusal. To this judgment I ought to have bowed, but then, on the other hand, the proprietors of *The Daily Graphic*, who, for the purposes of these letters, were my employers and who occupied the most favourable position for the formation of a practical opinion as to whether these letters did or did not displease the public, expressed to me very definitely and without qualification their satisfaction with the productions of which I was the author, but for which they were mainly responsible. A question of difficulty arises. Either the public read the letters, or it did not read them. If the public did not read the letters, then the proprietors of *The Daily Graphic* would have been dissatisfied at the results of an unremunerative outlay. But these gentlemen were not dissatisfied; therefore the public did read the letters. But the public only reads what it approves of, or what pleases it. Then I am led to a strange and terrible conclusion. Either the critics who condemned the letters were wrong, or, worse still, the public does not care twopence what the judgment of the critics may be. It is on the off-chance that this state of things, deduced by argument, may be the actual state of things that I again submit these letters to the public in another form. In the course of succeeding years many men and women will leave our shores to take up their abode in South Africa. Possibly some of these emigrants may glean from the following pages some information not altogether valueless as to the country, its people, its attractions, its modes of life and of travel. Moreover, of that large number of home-dwelling persons who follow with affection the fortunes of a great and growing colony in South Africa there may perchance be some whose interest therein may be quickened and sustained by the perusal of the experiences, the thoughts of an independent, unprejudiced wayfarer. In either case no harm is done; even a few grains of good may be produced.

Beyond mere verbal corrections and such other corrections as were necessary for the transposition of letters to a newspaper into chapters of a book, I have changed nothing of what I originally wrote, with the two following exceptions. Attempts at humour, or what is called "chaff," when taken seriously are failures so disastrous that they cannot be too quickly suppressed. Under this category come my allusions to the cook on board the *Grantully Castle* and my hazardous speculation on the origin of the female sex. This latter speculation, lightly turned off in a sentence, more for the purpose of an elegant termination to a letter than for the purpose of arousing controversy, was received so solemnly by grave and serious journals such as *The Spectator* and *The Speaker*, that they actually compared my ideas (unfavourably for me, I admit) with those of the illustrious Darwin. By the erasure of the guilty sentence alluded to from the text of these pages, I have done my utmost to withdraw from a competition so dangerous to myself.

I would add that the opinions which I expressed on the Dutch population of the Transvaal were intended by me to be exclusively confined to that population. Some imagined that those opinions were intended to apply generally to the Dutch in South Africa. But such wide and indiscriminate censure was far from my mind. The Dutch settlers in Cape Colony are as worthy of praise as their relatives, the Transvaal Boers, are of blame. The former, loyal, thrifty, industrious, hospitable, liberal, are and will, I trust, ever remain the backbone of our great colony at the Cape of Good Hope. That their numbers may increase, their influence develop, their possessions and their wealth expand, is my earnest hope, nor is it improbable that as time goes on the Dutch subjects of the Queen may communicate, by example and by intercourse, some of their excellent qualities to their backward brethren in the Transvaal. With these brief remarks, I submit to an indulgent public a narrative of a travel every hour of which was to me most enjoyable, a travel which I can confidently recommend to all who are desirous, and who are so fortunately situated as to be able, to make excursions for their pleasure into new parts of the world.

RANDOLPH S. CHURCHILL.

2, Connaught Place, W.,
  March 9th, 1892.

CONTENTS.

CHAPTER I.
OUTWARD BOUND.

Departure from Paddington Station—Reasons for the journey—The composition of the party—Arrival at Dartmouth—The *Grantully Castle*—Lisbon : the Zoological Gardens—Madeira—Invitation from Mr. Benet-Sanford—A sub-tropical garden—Farewell to Madeira—Shoals of flying fish—From breakfast to bedtime on board ship—Athletic sports at sea—Fire!—Cape Town 1

CHAPTER II.
CAPE COLONY.

Scenery and climate of Cape Town—Public Buildings—The Government House, Natural History Museum, and Public Library—Adderley Street—The sea prospect from Cape Town—Rivalry of Port Elizabeth—The inhabitants and environs of Cape Town—Dutch and English in the colony—Mr. Cecil Rhodes—The Transvaal War of 1881—Majuba Hill—Cape Politics—The South African States—Cape Town as a Coaling Station—Defences of the Cape—Forts at Simon's Bay—Fort Wynyard—General Cameron and the Cape Town Garrison 17

CHAPTER III.
DIAMONDS.

We leave Cape Town—The Paarl—Worcester Town—The

Hex River Pass—A Paddington man—Arrival at Matjesfontein—Mr. J. D. Logan—The Karroo—Diamond Industry at Kimberley—Visit to the offices of the De Beers Company—Mr. Cecil Rhodes a public man of the first order—Mr. Gardner Williams, mining engineer—The blue ground—Separating the diamonds—Precautions against Theft—The De Beers Company a model Village Community—Electric light used in the diamond mines 33

CHAPTER IV.
GOLD.

The diamonds of Kimberley—The journey to Johannesburg—Railway extensions—Grass veldt between Kimberley and Vryburg—The cattle farm of the future—"Native Reserve" of the Southern Bechuana—We reach Vryburg—Sir Sidney Shippard entertains us—Coaching with a team of mules—The wayfaring man in the Transvaal—An attractive little town—Gold mines in the neighbourhood—The outlook at Johannesburg—The gold mines—Selfish jealousy of the Boer Government—Astounding inequality of taxation—Bad condition of the roads to Johannesburg—The vicious system of concessions . 49

CHAPTER V.
MINING AND SPORTING.

Account of the Robinson Gold Mine at Johannesburg—The Langlaate Estate—Chlorination at the Ferreira Mine—Dr. Simon—The McArthur-Forrest process—Observations on the gold-fields of Johannesburg—Silver Mines in the Transvaal—Deer preserves—With Dog and Gun in search of Game . . . 65

CHAPTER VI.
THE TRANSVAAL BOERS.

A chance for British enterprise—The capacity of the

Transvaal and the incapacity of its rulers—The journey from Johannesburg to Pretoria—Description of Pretoria—The Dutch Parliament—From the Strangers' Gallery—An interview with President Kruger—Parliamentary manners—General Joubert—Report of a case showing the Boer idea of justice—Ill-treatment of Natives by the Boers—Shall we surrender Swaziland ?—The Withering Grasp of the Boer 79

CHAPTER VII.

ON THE ROAD TO MASHONALAND.

The Chartered Company's Station at Fort Tuli—Mining in the Zoutspanburg District—The Progress of the "Spider"—Our first cooking efforts—Hints for sportsmen—Sixty miles without water—A glimpse of Fairyland—We meet Major Sapte and Mr. Victor Morier—Meeting with Captain Laurie at Rhodes's Drift—The Bechuanaland Border Police—A "Boer trek"—President Kruger's position—Sir Frederick Carrington and the B.S.A.C. Co.'s police—Experiment with the new magazine Rifle 96

CHAPTER VIII.

THE EXPEDITION : ITS COMPOSITION AND EQUIPMENT.

Major Giles—A fine collection of giants—Our rifles and guns—Warning and advice to future travellers—Composition of the Expedition—Major Giles's trek from Vryburg to Tuli—The horse sickness in Africa—A camp fire concert at Fort Tuli 116

CHAPTER IX.

THROUGH BECHUANALAND.

Cold nights in camp—The horse sickness—Visit from Kaffir women to our Mariko River camp—Outspan on the banks of the Crocodile River—We cross the Mahalopsie River—Dr. Saur and Mr. Williams—Camp at Silika—Arrival at the Lotsani River—The luxury of a shave—The Suchi River—Headquarters of the Bechuanaland Police at Matlaputta—The Macloutsie River—I lose myself near the Semalali River while in quest of game—Catching up the waggons 126

CHAPTER X.

TREKKING AND HUNTING.

We entertain Sir Frederick Carrington—Farewell to Fort Tuli—The business of inspanning—Our camp at night—Sport with Dr. Rayner and Lee—Laying the telegraph wire—The Umzingwani River Camp—Koodoos, quaggas, and honey birds—Lee's boy nicknamed "The Baboon"—The elephant fruit-tree—Lee a charming companion on the Veldt—The Umsajbetsi River—Habits of our oxen and mules—Shooting game in South Africa—A native market—An unsuccessful antelope hunt—The mahogany tree—Further hunting experiences—Camp on the Bubjane River—Our conductor Myberg . . . 142

CHAPTER XI.

LIONS.

Lion Camp—The tales of a Huntsman—The snake-tree—In the track of the koodoos—We come across a posse of Lions—Antelopes and quaggas—Return to camp for the dogs—Result of one day's sport—We spend another day hunting—Provisions running short . 158

CHAPTER XII.

DIFFICULTIES OF TRAVEL ON THE VELDT.

The wealth of Mashonaland—We make a speedy trek and

overtake our waggons—Further losses by horse sickness—Stuck fast in Wanetse River—The Sugar Loaf and other miniature mountains—A pestilential spot on the Lundi River banks—A word of warning—Viandt, the Boer ostrich hunter—We reach Fern Spruit—Death of my shooting pony "Charlie"—A veldt fire—A day of discomfort and disaster—Providence Gorge—Description of Fort Victoria—Great loss of horses—Advice to intending emigrants . . 175

CHAPTER XIII.

CHARACTER OF THE COUNTRY BETWEEN FORTS VICTORIA AND SALISBURY.

Departure for Fort Salisbury—Our native workmen—Water in the desert—A dreary journey—The country between Fort Victoria and Fort Charter—Where is the 'Promised Land'?—We meet Mr. Colquhoun—The garrison of Fort Charter—From Fort Charter to Fort Salisbury—Lions in the way—The Settlement at Fort Salisbury—Signs of civilization—The gold districts of Manica, Mazoe River, and Hartley Hill—Reconnoitring after Game 193

CHAPTER XIV.

SPORT IN MASHONALAND.

Sport in South Africa—Hints to inexperienced sportsmen—Approximate cost of equipment for a six months' hunting expedition—Sir John Willoughby arrives at our camp on the Hunyani River—Hunting the Hartebeest—How to cook venison—A Slough of Despond—Further hunting adventures after antelopes—A native hunting party—A cobra in the camp—Method of scaring vultures off dead game—Accident to Major Giles—Scarcity of grain and food in Mashonaland—Return to Fort Salisbury . . . 212

CHAPTER XV.

THE GOLD DISTRICT OF THE MAZOE RIVER.

In quest of gold—Exploration syndicates—Mashonaland as a field for emigration—The Mazoe gold-fields—Captain Williams's report—Old workings—The "Golden Quarry" mine—Other mines visited in the district—More disappointments 234

CHAPTER XVI.

HUNTING THE ANTELOPE ON THE HIGH VELDT.

We start for Hartley Hill—The Mashonas as servants—Marriage in Mashonaland—All alone on the Veldt—Hints to hunters when lost on the Veldt—A Kaffir kraal—Barter with the natives—Dangerously bad shooting—The troubles of trekking—The country between Fort Salisbury and Hartley Hill—Wild flowers and fruit—Unsuccessful chase after ostriches—A fine herd of eland—The bull of the herd falls to my gun 246

CHAPTER XVII.

WEALTH OF MASHONALAND—DOUBT AND DISAPPOINTMENT.

Hartley Hill—Our party again united—The Tsetse-fly pest—Mr. Perkins joins me in a day's shooting—Surgeon Rayner's adventure with a lion—Contemplating the return journey—Making a clean breast of it—Deceptive appearances—Reefs in the Eiffel district—What is to become of the country?—Mr. Perkins and the leopard 263

CHAPTER XVIII.

LIFE AT FORT SALISBURY.

Mineral wealth of Mashonaland—Reefs in the Mazoe

River Valley—The "Matchless" Mine—Good news from Fort Victoria—A personal statement—Enterprise at Fort Salisbury—A model Ranche—Farms leased by the Chartered Company—An interesting auction—Indignation meeting against the Chartered Company—Horse-racing at Fort Salisbury—Organizing the administration of Mashonaland—Mr. Cecil Rhodes's views of the country 276

CHAPTER XIX.

ON THE ROAD HOME.

Second visit to the mines in the Mazoe Valley—Good-bye to Fort Salisbury—Bad roads—The officials of the Chartered Company—Fort Victoria once more—Climate and weather in Mashonaland—Gold discoveries round Fort Victoria—My faithful savage "Tiriki"—We telegraph home from Fort Victoria—Long's Mine—The Lundi River—Bad roads again—Death of a "salted horse"—The journey to Fort Tuli a record "trek" 295

CHAPTER XX.

LOOKING BACK.

Our method of travelling—Welcome and entertainment by the Bechuanaland Border Police at Macloutsie—Palapye, the capital town of Chief Khama—Lobengula, King of the Matabele—Meditated flight of all his tribe and belongings—The Bechuanaland Exploration Company—Conversation with Khama, Paramount Chief in the Protectorate—Palla Camp—The Journey to Mafeking—With Mr. Rhodes at Kimberley—The Agricultural and Mineral Resources of the Transvaal—My advice to young Englishmen 313

INDEX 331

LIST OF ILLUSTRATIONS.

FULL PAGE.

	PAGE
The R.M.S. *Grantully Castle*, 3489 tons, in Dartmouth Harbour	5
Luxurious Travelling in Madeira	9
A Madeira Caro, or covered Sledge	9
Parliament House, Cape Town	18
Government House and Gardens, Cape Town	18
Adderley Street, Cape Town	19
On the Road from Johannesburg to Pretoria.—Crossing a flooded river	54
The Market Place, Johannesburg	57
A Street in Johannesburg	58
"The Spider"	100
Executive Officers of the Expedition	117
A Camp Fire Concert at Fort Tuli	124
Showing a flare up for the lost one	140
The Members of the Expedition	142
First night out from Fort Tuli	144
Marketing with the Makalaka	153
Crossing the Lundi River	181
Two Members of the Expedition crossing the Lundi River	181
A Dreary Road.—The View fifty miles from Fort Charter	197
Building a "Scherm" to keep off Lions from the Cattle on the Hunyani River	204
A Sketch of the Country from Matipi's Kraal	204
Drawing dead Game home on a sledge made from the fork of a tree	229

LIST OF ILLUSTRATIONS.

	PAGE
On the Outskirts of Fort Salisbury	281
Nearing the end.—The Sale of the Surplus Stock and Stores of the Expedition at Fort Salisbury	286
Fort Salisbury.—At the Dentist's	295
The Arrival of the Telegraph Line at Fort Victoria.—Sending a telegram to London	305
From Tuli to Macloutsie	313
Crossing the Notwani after the heavy rains	325

TEXT ILLUSTRATIONS.

A Cape Cart	20
The Defences of the Cape.—A 9·2-inch breech-loading gun	29
£52,000 worth of diamonds classified for shipment at Kimberley	37
In the Rock Shaft of the De Beers Diamond Mine, at a depth of 900 feet	40
In the 800 feet level of the De Beers Diamond Mine	43
Sorting Gravel for Diamonds at Kimberley	44
General View of the Robinson Gold Mines at Johannesburg	66
Sir Frederick Carrington and Officers of the Bechuanaland Border Police and British South African Company's Police	106
Lord Randolph discussing his route with Sir F. Carrington at Fort Tuli	111
The long and the short of it	117
Camp Life at Tuli—Branding Cattle	121
Fording a River	130
The Main Column encamped on the bank of the Lotsani	133
The Camp of the Main Column at Suchi River	135
The Waggon Conductor sports a new pair of "store" trousers	138
Our Camp on the Umzingwani River	146
Typical Natives from the Umshlane River Districts	151

LIST OF ILLUSTRATIONS.

	PAGE
The "Sugar Loaf" Mountain between the Rivers Wanetse and Lundi	180
Passages in the Life of one of our Boys—In the Pantry	185
A "Veldt" Fire	187
One of our Boys (as he appeared with all his household goods)	194
One of our Boys (in sackcloth, drawing water)	195
The Camp before Fort Charter	200
Summer Sleighing in Mashonaland on the high road during the rainy season	**201**
Native Paintings on Rocks at Matefi's Kraal	**205**
Mr. Perkins, the Mining Expert, on the War-path	210
Visit to the Mazoe Gold-fields.—Experts at work	235
The Mining Settlement at Hartley Hill	264
At Hartley Hill.—Panning for Gold at Mr. Borrow's hut	272
Messrs. Johnson, Heaney, and Borrow's Ranche at Fort Salisbury	283
A Restaurant at Fort Salisbury	285
The First Horse-race at Fort Salisbury	288
A Party at the mess table, after dinner—Fort Salisbury	293
Tiriki	303
As he arrived	304
As he departed	305
The Outspan on the Tokwe River	**307**

ROUTE MAP . . . . . *At end of book.*

MEN, MINES, AND ANIMALS IN SOUTH AFRICA.

CHAPTER I.

OUTWARD BOUND.

Departure from Paddington Station—Reasons for the journey—The composition of the party—Arrival at Dartmouth—The *Grantully Castle*—Lisbon: the Zoological Gardens—Madeira—Invitation from Mr. Benett-Stanford—A Sub-tropical garden—Farewell to Madeira—Shoals of flying fish—From breakfast to bedtime on board ship—Athletic sports at sea—Fire !—Cape Town.

A BRIGHT morning towards the end of April. The eternal east wind blowing sharp and strong serves to moderate the regret which might be felt by one leaving England for a considerable period. In Paddington Station, alongside the platform, is drawn up the special express for Dartmouth. Every carriage appears to be full, round each compartment door large groups of persons, who intend to stay at home, wish farewell to those who are resolved to depart, and by their exuberant emotions obstruct the passage of the officials, of the tardy traveller, and of heavy trucks of baggage. Among the passengers the male sex largely predominates, and youth is stamped upon the countenances of the majority. In such a scene and in such a crowd I find myself an interested and active participator, for I, with a few friends, am starting on a long journey; and, in common with the others in the special train, my destination is South Africa. It happened to me shortly after my return from Egypt, in February, to meet Sir Henry Loch and Mr. Cecil Rhodes, the Governor and Prime Minister of Cape Colony, who had just arrived in England on a special mission of importance to the Home Government. Conversation naturally was mainly about South Africa, about the territories of the Chartered Company, the goldfields of Zambesia, the dispute with the Portuguese. Sir Henry Loch and Mr. Rhodes were kind enough to give me a cordial invitation to visit the Cape, and it suddenly occurred to me that I had really for the moment nothing better to do. Politics for the time attracted me little. The principal measure [1] before the House of Commons which was being pressed forward by the Government, and by the party to which I belong, I disliked intensely, and while I was not prepared to take part in any opposition to the measure, for motives which friends will appreciate, I was resolved to give no vote and say no word in its favour. A shareholder in the Chartered Company, and on intimate terms with some of the directors, my attention had already been turned to Mashonaland; I had imagined that the exploration and development of that vast country, so wealthy by

---

[1] "Irish Land Question Bill."

rumour and repute, was not unlikely to distinguish the close of the century. English and foreign interests had been and were clashing; spheres of influence for respective European Powers had been marked off in a hap-hazard and lighthearted manner; knowledge of the soil, of the climate, of the inhabitants, of the resources of Africa to the south of the Equator, was slight and by no means diffused among our people at home; I thought that the day might not be distant when it might be useful and beneficial that a member of Parliament might be able to offer to the House of Commons observations, opinions, and arguments based upon personal inspection, actual experience of those localities, and to place before the public the views and desires of Cape Colonists of authority and of Afrikanders generally which might have been personally confided to him. The attractions of travel, of the chase, and specially of seeking for gold oneself, of acquiring gold mines or shares in gold mines, contributed also to decide me on the enterprise, and the few weeks before departure had been well occupied with the somewhat elaborate preparations necessary for the journey and with business arrangements with friends who were similarly interested and attracted.

In the composition of my party I was fortunate. Major George Giles, late of the Royal Artillery, of considerable South African experience, who had seen much military service in that country, had undertaken to act as manager of the travelling and director of the route, and had preceded me by nearly a month to the Cape, intending to purchase at Kimberley the waggons, mules, oxen, and horses, to engage the men necessary for such an expedition. To travel in ox or mule waggons without greater discomfort and hardship than is incidental to camp life, a thousand miles to Mashonaland, several hundred miles exploring that country, a thousand miles return journey, occupying in the operation a period of not less than six months, requires a careful and costly collection of resources and plant, of which I will give a full description in a subsequent letter; to avoid undue delay while the season was favourable, Major Giles had gone on ahead of me to the Cape. My actual travelling companions at the moment of departure were Captain G. Williams, late of the Royal Horse Guards, who had amiably consented to assist me in my business and my writing, Mr. Henry Cleveland Perkins, an American mining engineer of great eminence, and Surgeon Hugh Rayner, of the Grenadier Guards, on leave, who intended to co-operate with the finest climate in the world in keeping us in good health, and to mitigate, so far as science might, the consequences of any accident or disaster which an untoward fate might inflict upon us. He was also instructed by the military authorities to furnish them with a report on the climatic and hygienic conditions of Mashonaland, to discover and specify healthy sites for camps, and to collect such other information as would be useful to possess beforehand, should military operations ever become necessary in that country.

To return to our departure: the last good-byes have been uttered, the doors of the carriages are slammed, the whistle sounds, and off we go, soon developing that alarming broad-gauge rapidity for which the Great Western was remarkable, dashing through Berkshire, Wiltshire, and Somersetshire with a haste, a hurry which seemed quite unnecessary considering the immense distance of travel which lay before us, and the considerable period of time which we had to do it in. Dartmouth, so familiar to the yachtsman, is reached all too soon. There lies the *Grantully Castle*, well known to fame as the ship in which Mr. Gladstone sailed when he made his celebrated "periplus" round Great Britain. A good-looking ship, sitting gracefully on the water, but small to my eye, more accustomed to the giants of the P. and O., of the Cunard, and of the White Star Lines, in which, when I have on former occasions crossed the ocean, it has been my fortune to travel. But the weather appears to be set fair, the sun is bright and warm, the sea smooth, and in fine weather and calm water a little ship does as well as a big one. Soon recede, from many a longing and lingering eye on board, the beautiful harbour, the green Devonshire cliffs, and a calm and moonlit midnight sees us well off Ushant. The *Grantully Castle* found the Bay of Biscay in a humour of comparative moderation; not that it was by any means amiable or attractive, on the contrary, it quite sustained its morose and ungenial character. The captain, indeed, expressed the opinion that it was as smooth as a mill-pond, but most of the passengers thought this description extravagant, many of them very practically disagreed with it. Fifty-eight hours after leaving London we anchored in the Tagus, opposite Lisbon, at the early hour of four a.m. Having to wait until five in the afternoon for the London mail, we took advantage of the opportunity to visit an ancient and historic city. The principal attraction to the eye was its cleanliness; broad, well-paved, clean-swept streets, spacious squares, adorned with interesting monuments, an environment of forest and green hills, offer an aspect calculated at first to please the stranger. But a something or other, difficult to describe, warns one instinctively that Lisbon is a city the fame and traditions of which lie exclusively in the past, in all probability never to be revived. The inhabitants wear a sleepy, almost a dead-alive kind of look. I did not observe a single Portuguese in the streets who appeared to be in the smallest hurry. No cheerfulness animates their countenances, as is the case with the population of the southern Italian towns. With the exception of a few public buildings, the edifices and dwelling-houses are of a poor and unpretentious character. There is a total absence of attractive and well-filled shops. Coming away, one feels that one is glad to have seen Lisbon, for the reason that it will be unnecessary ever to go there again. A drive through the streets terminated with a visit to the Zoological Gardens, interesting for the quantity of wild and of garden flowers, presenting the most brilliant hues, and for a singularly unique collection of monkeys, among which three intelligent and engaging chimpanzees for a time arrested our attention. I would strongly recommend any traveller to Lisbon not to omit to inspect these gardens. It must, however, be added that the people of the city scarcely appreciate their merits, for the place, during our visit of more than an hour, was totally deserted. The Botanical Gardens, a visit to which want of time compelled us to forego, are also said to be of considerable excellence.

The afternoon drawing on, it became necessary to return to the ship, and on the arrival of the London mail, we again put to sea, shaping our course for Madeira. A strong head wind and sea encountering us at the mouth of the Tagus, sadly thinned the attendance at dinner. The night was rough, and the following day supremely disagreeable. A driving mist, a warm, clammy wind, and a heavy rolling sea depressed the spirits, and made all long for more southern latitudes. At dawn on the second day after leaving Lisbon the island of Terra Santa stood out finely on the starboard bow. A few hours' steaming brought into clear view the bold outlines and grassy slopes of Madeira, and the sunny bay of Funchal. I was fortunate in finding, on arrival here, a note from an old friend, Mr. Benett-Stanford, who owns perhaps the most beautiful villa in the island, inviting my friends and myself to pass the morning with him. On landing, a slight ascent brought us to the doors of the Quinta Vigia (Anglice: The house and garden of the watch-tower) and to one of the most lovely gardens I have ever set eyes on. Imagine the contents of the conservatories and greenhouses of Tring Park, of Waddesdon, of Blenheim, Chatsworth, or Floors castle, all concentrated into a small space of some two acres, and growing with apparent wildness in the open air. For accumulated variety and profusion of bloom, fragrance, and luxuriant foliage, drawn from every quarter of the world, this garden can have no rival, and I have seen many, at home and in foreign countries. The gateway is sheltered by two remarkable camphor-trees, the leaves of which, when crushed in the hand, give off an aromatic and spicy perfume. On the terrace, overlooking the sea, one perceives, on the right hand, an immense mass of blue blossom, borne by the "Jacaranda." On the left a similar mass of deep crimson blossom draws attention to a splendid "Schotia." Two large trees, by name "Grevillia," covered with yellow flowers, offer a startling but agreeable contrast. Many fine specimens of the "dragon-tree" and of the "umbrella-palm" are studded about the ground. The "Strelitzia reginæ," apparently a sort of banana, is imposing from its size and its feathering leaves. The "peacock-tree" (Poinciana pulcherrima), with its lovely blossom, attracts the hand of the wanton flower-gatherer, and the eye is in reality dazzled by the extraordinary, but not inharmonious profusion of the "rose-apple" tree,

A MADEIRA CARO, OR COVERED SLEDGE.

cinnamon, silver banana, and mango tree, of the "Olea fragrans" and Francisia bushes, of the quaint "bottle-brush" plant, with its crimson flower. The verandah of the villa is covered with "Bougainvillia," and with another creeper, novel to me, the "Combretum coccinium," offering a mass of scarlet bloom. For the purpose of strolling through the town, our kind host provided a couple of hammocks, with their attendant bearers in white canvas clothing, and a "caro," or covered sledge, comfortably fitted, drawn by bullocks. This vehicle, which I imagine is peculiar to Madeira, can be drawn by the sleek oxen with considerable speed and perfect smoothness over the well-paved streets, or rather paths, which intersect the town and ascend the mountains. The public gardens, the opera house, a visit to the British Consul, to Messrs. Blandy's well-known office, and to the comfortable English club, exhaust the small time at our disposal, and soon after noon we are once more on board the *Grantully Castle*, heading south to Cape de Verde, leaving fast, but somewhat reluctantly, Madeira behind us.

The hour of 5.30 a.m. finds two or three passengers on deck enjoying the pleasures of a cup of coffee and the first morning cigarette. From six to eight the ship's toilette proceeds, from the rough-and-ready washing and scrubbing of decks to the conscientious and minute polishing of every bit of wood and brass-work. This is, perhaps, the pleasantest time of the day; cool, fresh air, peaceful decks unobstructed by chairs; walking exercise is possible and pleasant, no noise disturbs the current of your morning reflections. Cloudless is the sky, strong the following "N.E. trade," covering the surface of the sea with white horses, but not strong enough to overtake and refresh the fast-flying ship, whose passengers are warned by the already glowing sun that the day is going to be a "piper." Looking over the stern you see the same troop of swifts or swallows which for three days now have followed us. Swooping, hovering, skimming, darting, never left behind, seemingly never progressing, never resting; where they come from, where they are going to, where they sleep, and what they feed on offers a problem which natural history has not yet attempted to solve. From time to time shoals of flying fish shoot from the waves, nor can one imagine a more pleasant sight than these animated particles of silver present, jumping from and skimming along the surface of the water by scores and hundreds, gleaming and glistening in the sunlight. Last evening, one, probably a "loose fish," when all his fellows had gone to bed, jumped with a mighty leap right on to our deck. Promptly secured by a quartermaster, it attracted the observation of a French gentleman, who desired to photograph it, have it stuffed, and carry it home to his family museum. But my friend and I disappointed him, taking it and giving it to the cook, and eating for breakfast the finny fowl. Morning wears away, groups of gentlemen have appeared and disappeared, clothed in that loose and light attire, with sponge, towel, and soap, which denotes resort to or accomplishment of the matutinal bath. Breakfast is over, and by noon most of the party are deep in literary occupation, writing and reading apparently being suitable only to this time of day. A temperature of 75 deg. under the awning is adverse to muscular exercise. Now come round the managers of the daily lottery on the ship's run, drawing variously from the passengers shillings and crowns; some contenting themselves with a single lot, others basing their hopes on securing many chances. At 12.30 the captain announces that in the preceding twenty-four hours we have compassed the respectable distance of 328 miles. The winner of the lucky number looks happy in his clever superiority, but can scarcely be much richer, for etiquette prescribes that he should generously proffer champagne to the losers. Luncheon at an end, the passenger mind turns to amusement. The young find relief in the violent exercise of deck-cricket, and in the wild mirth occasioned when some placid, reposing, and digesting person receives the ball full in his face, an incident which he is expected to bear with perfect equanimity, neither asking for, nor expecting the smallest apology. Deck-quoits, and the sonorous game of "Bull," claim their adherents; for others, the frivolous "Halma," the rattling backgammon, or the severely serious chess have their charms, and here and there vice betrays itself in the shape of cards, with its usual accompaniments of brandy and soda, beer, pipes, and cigars, while ever and anon a half-suppressed " damn," or a less suppressed exclamation of triumph, discloses the vicissitudes of the game. So the afternoon blends itself into the evening. At half-past six the bell for dinner sounds, after which, music in the saloon detains many. "Gentlemen oblige" the company with songs and recitations, the French gentleman displays a startling ingenuity in card conjuring. On deck, soothed by tobacco, many groups converse. The talk is probably of Africa and Mashonaland, of diamond mines and of the " Randt," of the depth and thickness of reefs, of the yield of so many pennyweights to the ton, of " paychutes," of stamps, crushers, and chlorination, till the disappearance one by one of the electric lights warns us that the night is well on, and we turn into our cabins to dream of re-discovering El Dorado, of revelling in the " placers " of another California, of handling deliriously the nuggets of a second Ballarat. On waking we trust that we have not dreamed of a Golden Fleece. So the voyage proceeds.

One of our days at sea was agreeably passed by holding athletic sports and contests, in which all classes of the passengers took part. Racing, leaping, and cock-fighting were the principal features. Four times round the ship from stern to bow was found to afford a good half-mile course, and the struggle was decided after several heats, some of them of an exciting character. The prize for leaping was long and closely contested, four feet seven being at length triumphantly cleared by the victor. Taking into account the perceptible rolling of the ship, the achievement appears to have been one of merit. Cock-fighting, also, was the source of considerable amusement, and, to the astonishment and delight of all, the prize was carried off by the smallest and youngest of the competitors. A potato race, an egg and spoon race, in which ladies only took part, an obstacle race, in which last suspended lifebuoys to be darted through and long wind sails to be crept through tested severely the agility and endurance of the runners, occupied fully the morning and the afternoon. A respectable fund for prizes had been previously collected, the proceedings were managed with the utmost order and method, the decisions of the officials received without a murmur. In the tug-of-war the first class overpulled the second, who also suffered defeat at the hands of the third-class passengers. It appears that these athletic sports are a regular institution on board the *Grantully Castle*, and this original and agreeable method of breaking the monotony of a long voyage offers an example to commanders of ocean-going vessels which cannot be too widely imitated. The day antecedent to our arrival at Cape Town was the most disagreeable of the voyage. A high and heavy rolling sea rendered sleep by night or occupation by day alike almost impossible. Shortly after luncheon an incident occurred which for some moments must have fluttered the strongest nerves. A strong smell of burning, smoke coming up thickly from the after skylight, passengers running up from below, driven out of their cabins by the stifling smoke, told us all too plainly that a fire had broken out on board ship. The fire-bell was rung, the officers and crew were beat to quarters, the hose was fitted, and in a few minutes gallons of water were being poured through the skylight down into the after-hold, where it was discovered the fire had occurred. At the same time rapid preparations were made for getting the boats ready for lowering, though whether these would have been of much service to us, had we had to have recourse to them in such a heavy sea, was a matter of serious doubt. In a quarter of an hour, however, or twenty minutes, all danger was over. The officers and crew worked with the utmost steadiness and resolution, the first officer particularly showing extreme courage and endurance, being the first to descend into the burning hold, and remaining in an almost impossible atmosphere for a considerable time directing the application of the water. The passengers preserved their composure remarkably, contenting themselves with getting out of the way, and offering as little impediment as possible to the operations of the ship's company. The cause of the fire was not discovered to a certainty. In the after-hold were scattered a variety of ship's stores, a quantity of empty bottles, heaps of straw and shavings, the contents of opened packing-cases. It was ventilated by a grating into the cabin passage, and the captain supposed, probably with justice, that some reckless and wanton passenger, lighting a cigar below in violation of rigid rules, had ignorantly and carelessly thrown away the match still lighted, allowing it to fall on this mass of inflammable material. The mail-room adjoins the afterhold, and the mails must have had a narrow escape; while the hold immediately forward contained large stores of spirits. Had these been ignited the consequences would probably have been most serious. Some inches of water on the cabin floors, and a strong smell of smoke, of charred straw and wood, served for some hours to remind the passengers how near they had been to a very unpleasant termination of their voyage. The morning of the 14th of May broke gloomily, with heavy rain and driving mist. About nine o'clock a bright clearance to the southward disclosed the heights of Table Mountain. As we drew nearer the weather became brighter, the clouds broke: when off the breakwater the Cape of Good Hope was welcoming us with its sunniest smile. The approach from the sea to Cape Town is imposing and attractive. The lofty granite mass of Table Mountain, the distant ranges of hills stretching over half the horizon, and the calm waters of Table Bay brought into the mind successively Gibraltar, the Riviera, and the Bay of Palermo, while the attractions of the spot were strengthened by the feeling that a long, tedious, and monotonous voyage had at length been accomplished. It may be a matter of question whether, under present conditions, a voyage to South Africa is as beneficial to invalids or to persons of delicate health, and liable to sea-sickness, as is generally supposed. The excessive heat in the regions of the Equator debilitates and exhausts; scarcely a day of the voyage was not marked by considerable rolling or pitching, and the imperfect ventilation, the inferior food, and the want of power and speed in the older Cape vessels lead one to hope that before long an increasing volume of passenger traffic may compel the construction of larger, better found, and swifter ships, rivalling in their excellence the racers of the North Atlantic lines. I imagine that a vessel like the *Teutonic* could cover the distance between Plymouth and the Cape of Good Hope in less than fourteen days. The *Grantully Castle* occupied a period of nineteen days and nineteen hours. On landing, I repaired to Government House, to which I had received a gracious invitation.

CHAPTER II.

CAPE COLONY.

Scenery and climate of Cape Town—Public Buildings—The Government House, Natural History Museum, and Public Library—Adderley Street—The sea prospect from Cape Town—Rivalry of Port Elizabeth—The inhabitants and environs of Cape Town—Dutch and English in the Colony—Mr. Cecil Rhodes—The Transvaal War of 1881—Majuba Hill—Cape Politics—The South African States—Cape Town as a Coaling Station—Defences of the Cape—Forts at Simon's Bay—Fort Wynyard—General Cameron and the Cape Town Garrison.

> Ille terrarum mihi præter omnes
> Angulus ridet, ubi non Hymetto
> Mella decedunt viridique certat
> Bacca Venafro;
> Ver ubi longum tepidasque præbet
> Jupiter brumas, et amicus Aulon
> Fertili Baccho nimium Falernis
> Invidet uvis.

FOR beauty of scenery and general excellence of climate Cape Town approaches perfection. Inhabited by some 50,000 souls, it reposes at the foot of the great Table Mountain, sheltered though not oppressed by towering and precipitous granite masses. Possessing and proud of a history going back over a period of upwards of 250 years, the town itself shows few if any signs of antiquity. The traveller might often imagine from its straggling and unfinished appearance that he had arrived at one of those sudden settlements, the creation of a few months or weeks, which are characteristic of an American territory or of the Australian bush. Public buildings of high architectural merit are scarce; indeed, the Houses of Parliament and the Standard Bank may be said to be the only edifices entirely worthy of the traditions and position of the town. The old Town House, the old Castle, vividly and agreeably recall the Dutchman of the seventeenth century, relics of an interesting past, testimonies of a famous history, which should be tenderly preserved. His Excellency the Governor is respectably, but not splendidly, accommodated. A long, low building, hidden away in a corner of ugly elevation but of roomy and commodious interior, containing spacious apartments, uneasily supports the dignified title of "Government House." A garden of considerable extent, well filled with shady oak-trees and many fine specimens of tropical plants, makes up largely for the architectural shortcomings of the edifice. Adjoining are to be found the Botanical Gardens, the Natural History Museum, and the Public Library. All of these institutions apparently suffer from a want of liberal maintenance, which is the more to be regretted as their contents are for the most part excellent and rare. The Natural History Museum possesses a very perfect collection of African fauna, mainly contributed by the famous hunter, Mr. Selous, of birds, and of mineralogical and conchological specimens of great interest; but

PARLIAMENT HOUSE, CAPE TOWN.

GOVERNMENT HOUSE AND GARDENS, CAPE TOWN.

ADDERLEY STREET CAPE TOWN.

all these objects of study are so crowded and so crammed up together, and stowed away in cases so insufficiently lighted, that detailed and careful inspection of them is a matter of extreme difficulty. Both this museum and the Public Library, which latter is a fine hall, containing a large and varied collection of books and many ancient manuscripts, are freely and frequently resorted to by the inhabitants. Adderley Street in the morning is crowded and animated; many of its buildings have striven to attain to a respectable standard of civic architecture, and well-filled shops elegantly display a variety of articles of luxury, which suggest the diffusion of an easy affluence. The sea prospect from Cape Town is most agreeable. A lengthy breakwater, constructed with great solidity, protects an anchorage where many vessels of size might congregate. The docks, which were large enough for the shipping requirements of a generation ago, are too small to allow of the entrance of large modern steamers, and it is much to be regretted that the construction of a big, wide dock in the rear of the existing docks has been suspended owning to want of funds. Port Elizabeth, with superior railway advantages, and, perhaps, a more go-ahead public spirit, is pressing Cape Town hard; and it is probable that, if the latter does not bestir itself, it may forfeit its commercial eminence in South Africa. But possibly the charm of Cape Town lies in its respectable repose. The inhabitants, who welcome the stranger with a cordial hospitality rarely to be found elsewhere, have inherited, probably from the Dutch, a pleasant conservatism of thought and of habitude. They are not inclined to believe that the bustle of Melbourne or the crowds of Sydney represent the highest standard of social happiness; they have a tendency to regard with some doubt and anxiety the development and progress which Cape Town

A Cape Cart.

has undoubtedly made in the last few years; many of them view with apprehension and some with alarm the influx of a large population which may shortly be attracted by the mineral wealth of South Africa already said to be discovered. The late Lord Iddesleigh, in one of his political discourses, averred that he had been accused of being "wanting in go," but it was felt by all that the accusation, if true, only exhibited more pleasantly the general amiability of his character. Similarly it is possible that the people of Cape Town have a tendency to a liability to such an accusation, but those who are fortunate enough to know and understand them will readily confess that the defect, if it exists, may be counted among their attractions rather than among their faults. The environs of Cape Town in the direction of Wynberg are of surpassing beauty. Forests, groves, plantations of oak, pine, eucalyptus, owing their origin to the provident forethought of the early Dutch settlers, thickly cover the ground from the slopes of the mountain almost to the shores of the sea. Miles of shady lanes, extending in all directions, make riding and driving an unfailing pleasure, while on every side old-fashioned villas and country-houses, with perfect and well-kept gardens, disclose alike the cultivated taste and the love of country life which characterize the wealthier portion of the resident community. English people afflicted at home by a winter climate which year after year grows more intolerable and more interminable, fruitlessly, and at great cost, seek sunshine and warmth in the south of Europe amid unsympathetic foreigners. A three weeks' voyage, unaccompanied either by hardships or risk, would bring them to this lovely spot, where, among people of their own race, speaking their own language, and thinking their own thoughts, they would find and enjoy the most temperate and equable summer weather, with all the attractions of sea-side existence which the earth can offer.

Socially a very happy change has, in recent years, been effected in the Cape Town community. The old hostility between the English and the Dutch, which at the time of the Transvaal War had attained a dangerous height, seems to have entirely passed away. The two sections regard each other with feelings of respect, friendship, mutual trust. The genius of the Prime Minister, Mr. Cecil Rhodes, has mainly contributed to this auspicious state of things. He has known how to acquire and retain the confidence of the English and of the Dutch colonist, he has shown them in the daily practice of his Government that their interests are entirely and absolutely common, and so homogeneous is now this Cape community that the President of the South African Republic and the Transvaal Boers have been plainly and effectively warned by many Dutchmen of authority and position in Cape Colony that unfriendly action on their part against the British position in Zambesia, and hostile action by Boer "trekkers," against the British Chartered South African Company, will neither receive the support nor enjoy the sympathy of any appreciable section of the Dutch subjects of the Queen. The Cape Colony Dutch sympathized profoundly with their countrymen, who, in 1881, were fighting for their freedom; but that freedom having been restored and guaranteed, they are equally ready to disapprove of, and even to resist, their Transvaal kinsmen impelled by land hunger or by sheer animosity to attack British possessions and British subjects without reason or provocation. Moreover, the Cape Colony Dutch argue with much force: "We supported you Boers in your struggle for liberty, our support saved you from British resentment; in return you have placed prohibitive duties on our goods and productions, you have obstinately hindered the extension of our railways, and you have excluded our children from civil employment in your State. Whereas we find that this Imperial Goverment which you so unreasonably hate, wether in Bechuanaland or in the Chartered territory, admits our goods duty free, actively supports the development of the railway system, and invites our children not only to enter its service, but to come into and occupy the lands under its control." In justice it should be added that the sagacious policy of Mr. Rhodes has only been made possible by the termination of the Transvaal War in 1881, and by the manner of its termination. The surrender of the Transvaal and the peace concluded by Mr. Gladstone with the victors of Majuba Hill were at the time, and still are, the object of sharp criticism and bitter denunciation from many politicians at home, *quorum pars parva fui*. Better and more precise information, combined with cool reflection, leads me to the conclusion that, had the British Government of that day taken advantage of its strong military position, and annihilated, as it could easily have done, the Boer forces, it would indeed have regained the Transvaal, but it might have lost Cape

Colony. The Dutch sentiment in the Colony had been so exasperated by what it considered to be the unjust, faithless, and arbitrary policy pursued towards the free Dutchmen of the Transvaal by Sir Bartle Frere, Sir Theophilus Shepstone, and Sir Owen Lanyon, that the final triumph of the British arms mainly by brute force would have permanently and hopelessly alienated it from Great Britain; Parliamentary government in a country where the Dutch control the Parliament would have become impossible, and without Parliamentary government, Cape Colony would be ungovernable. The actual magnanimity of the peace with the Boers concluded by Mr. Gladstone's Ministry after two humiliating military reverses suffered by the arms under their control became plainly apparent to the just and sensible mind of the Dutch Cape Colonist, atoned for much of past grievance, and demonstrated the total absence in the English mind of any hostility or unfriendliness to the Dutch race. Concord between Dutch and English in the colony from that moment became possible, and that concord the government of Mr. Rhodes inaugurated, and has since to all appearance firmly riveted. On the other hand, the peace thus concluded with the Transvaal carried with it some grave disadvantages. The re-erection of the South African Republic contributed another powerful factor to the forces of disunion in South Africa; the Boers of the Transvaal, wanting altogether the common-sense of their kinsmen in the colony, have since the war been inflated with an overweening pride, foolishly eager to seek quarrels and sustain disputes with the English power, and will continue, possibly for generations, to be a formidable obstacle to either political or commercial federation in South Africa. Moreover, the generosity of the surrender of the Transvaal by the English Government was naturally misunderstood by, or was not apparent to, the mind of powerful native races. On the whole, I find myself free to confess, and without reluctance to admit, that the English escaped from a wretched and discreditable muddle, not without harm and damage, but probably in the best possible manner, and that lessons have been taught to many parties by the Transvaal war which, if learned, may be of the utmost value in framing future policy.

South African politics are highly interesting at the present moment. The position of the Cape Government is one of apparent solidity and power. Against it, supported as it is by a preponderating majority in Parliament, two ex-Prime Ministers, in imperfect harmony with each other, and followed by groups numerically insignificant, with difficulty sustain the forms of an Opposition. Complete concord and co-operation exist between the Parliament and the Ministers on the one hand, and the High Commissioner on the other. It is, indeed, well that this should be so now, for the development of the agricultural and mineral resources of Matabeleland under the protection of the British Government, through the instrumentality of the

Chartered Company, will require for years the most skilful, prudent, and courageous handling. That those resources will before long prove to be of value to the English people does not admit of doubt; but their very value excites the cupidity, not only of the weak and easily-controlled Boer, and of the weaker and still more easily-controlled Portuguese, but also of such powerful rivals as France and Germany; any failure on our part effectively to develop Matabeleland, to preserve peace, order, and security in those vast regions, and to combine in the work the entire British African community, would result in a loss which, from a national and from a commercial point of view, can only be described as immeasurable. But the combination of the British African community for effective executive purposes is a task which may almost exhaust the resources of statesmanship. From the Zambesi to the Cape of Good Hope, a region occupying some two thousand miles of land in length, inhabited by about half a million whites and by over four millions of natives, every form of government known to history is to be found in existence and at work: in the Cape Colony a representative Parliament elected on the widest native and European suffrage, with responsible Ministers and almost complete independence of the Home Government; in Natal a more restricted representative body, with Ministers not directly responsible to that body, a sort of Prussian administration; in Zululand the personal and direct government of the Governor of Natal; in the

Transvaal an independent republic, **but unable to conclude treaties with** foreign States **without the approval of the British Government; with a president, executive, and** two chambers **elected by** Dutch burghers, but with many thousands of European population possessing no political rights; in the Orange Free State another independent republic, governed by a president and one chamber, elected by all the citizens; **in Bechuanaland the** direct and personal **government** of the Governor of Cape Colony, **exercised through an administrator** under laws enacted **by the Governor's proclamation**; in Basutoland direct and personal **government** of the High Commissioner, exercised through an administrator **under** laws enacted by the High Commissioner's **proclamation, and with native** customs and **native laws** administered by native chiefs so far as they may not be inconsistent with English justice; in the Bechuanaland Protectorate the personal authority of the High Commissioner, the native chiefs and territory protected by the Bechuanaland Border Police from external aggression, with **native laws** administered **by the** chiefs; in Swaziland a joint **Government carried on by the** British **and the Transvaal authorities,** with an executive consisting of a representative of the Swazi nation, of the British Government, and of the South African Republic, whose **laws require** the joint approval of the guaranteeing Powers; in Pondoland an independent native State, the sea-coast only of which is under the protection of England with **a** seaport **on** the coast which is actual British territory; in Damaraland a German territory under direct control of the German Government, the only seaport of which belongs to the Cape Colony; in Amatongaland an independent native State governed by Queen Zambili and her councillors, but possessing no power to conclude treaties with foreign States, except with the approval of the British Government; while in Matabeleland, a territory as large as France, inhabited by a numerous and warlike tribe of Zulu origin, the burden of government has been assumed by an English Commercial Corporation under charter from the Crown, under the jurisdiction of the High Commissioner, and deriving administrative authority from the High Commissioner acting on behalf of the British Government. The mere enumeration of these various forms of government, the mere setting out of this complicated and variegated congeries of powers and authorities all mixed up almost inextricably together, will suffice to give some idea of the difficulties and embarrassments which attend the course whether of a Secretary of State, of a High Commissioner, or of a Cape Colony Government and Parliament.

For a British coaling station of great importance Cape Town is fairly well fortified. By the courtesy of General Cameron, commanding the forces, and of Colonel Knollys, commanding the Artillery, I was enabled to make a detailed examination of the defences. I was naturally much interested in noting what had been done, because

Cape Town is one of the coaling stations which Lord Salisbury accused me of being desirous, when Chancellor of the Exchequer in 1886, of leaving in a defenceless condition. Five 9·2-inch breech-loading guns, with hydro-pneumatic mountings, placed in positions selected with admirable art, make the approach of a hostile fleet a work of great difficulty and danger. With these formid-

The Defences of the Cape—A 9·2-inch breech-loading gun.

able cannon are placed, more for purposes of ornament than of effective use, fourteen or fifteen seven-ton muzzle-loading guns. These latter have been discarded by the navy, and are considered by expert artillerists, on account of their muzzle-loading arrangements, their inferior accuracy, and small penetrative power, to be obsolete. The Home Government, however, considered them to be good enough for the Cape, and at great expense have sent out and mounted a number of them for the defence of Table Bay and of Simon's Bay. I am informed that for the same money an equal number of the new six-inch breech-loading gun might have been furnished, in which case the defences of the Cape of Good Hope in respect of ordnance would have been complete.[1] But it is ever so. Our War Office and Admiralty can never be persuaded to make a finished and perfect work. The various detached forts in which these guns have been placed have been constructed with great solidity. The Cape Government supplied the sites and the labour at a cost of about 60,000*l.*; the guns and mountings were furnished by the Home Government. A 9·2-inch breech loading gun, with hydro-pneumatic mounting, costs 17,000*l*. Two of these guns, two 9-inch muzzle-loaders, together with several seven-ton muzzle-loaders, command and protect the naval station at Simon's Bay. The forts at Simon's Bay have been so ingeniously concealed by the engineers that it would be difficult and perhaps impossible for the officers of an approaching hostile fleet to discover their situation until it had come well within range. The General kindly allowed the 9·2-inch breech-loading gun in Fort Wynyard to have three rounds fired from it for my inspection. The projectile

---

[1] The facts set out above were controverted in the House of Commons, were controverted by Mr. Stanhope, Secretary of State for War. They are nevertheless authentic and accurately stated.

weighs 380lbs., and the charge of powder is 166lbs. The target, a flagstaff on a barrel, was moored at a distance out at sea of 2900 yards. All three shots, so excellent is the accuracy of the gun and the training of the gunners, passed within a few feet of the target, which, if it had been an enemy's ship, would have suffered fatal damage. The destructive range of this cannon attains the prodigious distance of 10,000 yards. The garrison at Cape Town is ridiculously weak. It consists of one and a half battalions of infantry and two batteries of artillery. The 9·2-inch breech-loading gun requires for its handling seventeen trained artillerists. The garrison in its present feeble state can only furnish three men per gun. A scheme is at present being considered for combining effectively with the Imperial troops the best volunteer forces, but nothing has been decided on in this direction, and the jealousy of home interference entertained by the Colonial Government may yet cause much friction and delay. I am informed that if Cape Town is to be rendered at all secure from hostile attack, one additional battalion of infantry and one battery of artillery is imperatively required. General Cameron, speaking at the Mayor's luncheon, asserted that he had only 1300 effective men under his command, whereas a garrison of at least 6000 was necessary for the adequate protection of the Cape of Good Hope. To do full justice to the War Office I should add that it maintains with magnificent liberality for the duties of this feeble garrison a staff equal in numbers and importance to the requirements of an army of 20,000 men.[1] With the due celebration of the Queen's Birthday on the 25th May, and with the opening of the Cape Parliament on the following day, at both of which interesting and imposing ceremonies I was privileged to be present, I brought to a close a most agreeable fortnight passed at Cape Town.

[1] This statement was also contradicted in the House of Commons by Mr. Stanhope, Secretary of State for War. It is, nevertheless, perfectly accurate.

CHAPTER III

DIAMONDS.

We leave Cape Town—The Paarl—Worcester Town—The Hex River Pass—A Paddington man—Arrival at Matjesfontein—Mr. J. D. Logan—The Karroo—Diamond Industry at Kimberley—Visit to the offices of the De Beers Company—Mr. Cecil Rhodes a public man of the first order—Mr. Gardner Williams, mining engineer—The blue ground—Separating the diamonds—Precautions against theft—The De Beers Company a model Village Community—Electric light used in the diamond mines.

THE mail train for Kimberley leaves Cape Town in the evening, but the traveller will be well advised in doing the portion of the journey as far as Matjesfontein by day. The railroad passes through mountain scenery of exceptional beauty and variety. We left Cape Town in pouring rain, but soon after our departure the clouds broke, and the sun shining out brightly upon the mist which hung over the tops of the Hottentot Holland Range produced a series of rare and astonishing rainbow effects. About thirty miles from Cape Town is situated the Paarl, an old town of French origin, which takes its name from a chain of large granite boulders, supposed to resemble the pearls of a necklace, which adorn the summit of the hill overhanging the place. The valley of the Paarl is covered with vineyards interspersed here and there with fertile pastures on which graze herds of long-horned cattle. The sweet wines of the Paarl rival in excellence those of Constancia. The slopes of the hills are well wooded, the general aspect of the valley is of extraordinary attraction. Round about Wellington, a town in the vicinity of the Paarl, we find a great corn-growing country, while in the division of Worcester, a hundred miles distant from Cape Town, we come to a district producing an abundance of sheep, cattle, horses, ostriches, corn, dried fruits, and wines. The town of Worcester has been admirably constructed. "The streets are well laid out in parallelograms, bordered with water-courses and eucalyptus trees. A plentiful supply of water, conveyed from the Hex River by an artificial canal six miles in length, imparts to the town an unusually verdant aspect, the charm of which is enhanced by contrast with the sterile appearance of the lofty and rugged mountains which surround it. About nine miles from Worcester there are hot springs with a temperature of 145 deg., the water of which is famed for its cure of wounds and skin diseases."[1] Speaking generally of the road between Cape Town and the summit of the Hex River Pass, the mountain and valley scenery is hardly to be surpassed in loveliness, and the traveller, assisted by memory or imagination, may fancy himself now in Yorkshire or Worcestershire, now in the Highlands of Scotland, now in the valleys of Provence. After leaving Worcester, the railway ascends the Hex

[1] *South African Argus Annual.*

River Pass, the summit of which is some 3000 feet above the level of the sea. The road, which is a triumph of engineering skill, has an average gradient of 1 in 40, and no fewer than seventy-two curves, some of them very sharp. To view the magnificent mountain prospect with more advantage our party was accommodated with seats on the little platform in front of the engine, and the sensations of travelling in this manner along the verge of precipices of giddy depth and over iron bridges of frail appearance were at once novel and thrilling. The engine-driver, who was very affectionate, confided to me while we were passing at respectable speed an apparently dangerous portion of the track, that he was a Paddington man. This is the second gentleman occupying an official position connected with the administration of the railroads of South Africa who claimed Paddington as his birthplace and his home, and who saluted with joy the appearance of his representative in Parliament.

Matjesfontein, which we reached in the evening and where we passed the night, affords a remarkable example of what can be effected by the energy of the Englishman. Mr. J. D. Logan, who is the proprietor of an estate here of 100,000 acres in extent, has settled himself down on what appears at first sight to be the most unpromising spot for a farmer which the mind can imagine. Here, in the arid plain of the Karroo, producing nothing but low scrub and scanty herbage, he has built himself a large and comfortable house, a spacious homestead with good cottages for his men, and elevates with much success flocks of many thousands of sheep and herds of many hundreds of cattle. The Karroo is far more hospitable and nourishing for live stock than the uninstructed tourist would imagine. The climate is perfect, the air invigorating like that of Scotland, and the only source of anxiety to the farmer is found in the somewhat insufficient rainfall. Sport is to be obtained in plenty by the sportsman who does not fear hard work. The quail, the namaqua partridge, the koran, the pauw, a species of bustard, all at different times and seasons fill the game-bag, while often the spring-bok and sometimes the leopard will fall to the well-aimed rifle. I imagine that many a young English farmer with a good training, an active disposition, and a small capital, might find in the Karroo both a home and a fortune. No rent, scarcely any taxes, and perfect freedom are constituents of happiness which to the ordinary English farmer would appear almost as an unrealizable dream. An early start the next morning, a journey of twenty-four hours across the monotonous and apparently limitless expanse of the Karroo brought us to Kimberley. Nothing in the external appearance of Kimberley suggests either its fame or its wealth. A straggling, haphazard connection of small, low dwellings, constructed almost entirely of corrugated iron or of wood, laid out with hardly any attempt at regularity, and without the slightest trace of municipal magnificence, is the home of the diamond industry. It seems that when the diamonds were first discovered some twenty years ago, many thousands of persons settled down suddenly on the spot like a cluster of swarming bees, and established themselves anyhow as best they could in the most rough and primitive fashion, never dreaming but that the yield of diamonds would be of limited extent and of short duration, that their fortunes would be rapidly acquired, and that they would pass as rapidly away from the

Classified for shipment at Kimberley.

place, having exhausted all its wealth-producing resources. The reverse has proved to be the case. The diamondiferous resources of Kimberley are now known to be practically inexhaustible, but the amalgamation of the mines has restricted employment and checked immigration, and the town still preserves, and probably will always preserve, its transitory and rough-and-ready appearance. There are, however, a number of excellent shops, and there are few articles of necessity, of convenience, or of luxury which cannot here be purchased. A most comfortable and hospitable club, an admirably laid-out and well-arranged racecourse testify to the thoroughly English character of the settlement. At Kimberley the diamond is everything, and the source and method of its production claim more than a passing mention. My first visit was to the offices of the De Beers Company, which company represents the amalgamated interests of the De Beers, Kimberley, Bultfontein, Du Toits Pan, and other smaller mines. The amalgamation was the work of Mr. Cecil Rhodes, and it was this great work, accomplished in the teeth of unheard-of difficulties and almost insurmountable opposition, representing the conciliation and unification of almost innumerable rival jarring and conflicting interests, which revealed to South Africa that it possessed a public man of the first order. The scale of the company's operations is stupendous. On a capital of nearly 8,000,000*l.* of debenture and share stock it has paid, since its formation in 1888 up to March, 1890, interest at the rate of $5\frac{1}{2}$ per cent., and an annual dividend of 20 per cent. In the same period it has given out some two million five hundred thousand carats of diamonds, realizing by sale over three and a half million pounds, produced by washing some two million seven hundred thousand loads of blue ground. Each load represents three-quarters of a ton, and costs in extracting about 8s. 10d. per load, realizing a profit of 20s. to 30s. per carat sold. The annual amount of money paid away in interest and dividends exceeds 1,300,000l. The dividends might have been much larger, but the policy of the present Board of Directors appears to be to restrict the production of diamonds to the quantity the world can easily absorb, to maintain the price of the diamonds at a fair level from 28s. to 32s. per carat, and, in order the better to carry out this policy, to accumulate a very large cash reserve. I believe that the reserve already accumulated amounts to nearly a million, and that this amount is to be doubled in the course of the next year or two, when the board will feel that they have occupied for their shareholders a position unassailable by any of the changes and chances of commerce. In the working of the mine there are employed about 1300 Europeans and 5700 natives. The wages paid range high, and figures concerning them may interest the English artisan. Mechanics and engine-drivers receive from 6l. to 7l. per week, miners from 5l. to 6l., guards and tally-men from 4l. to 5l.; natives in the underground works are paid from 4s. to 5s. per day. In the work on the "floors," which is all surface work, overseers receive from 3l. 12s. to 4l. 2s., machine men and assorters from 5l. to 6l., and ordinary native labourers from 17s. 6d. to 21s. per week. In addition, every employé on the "floors" has a percentage on the value of diamonds found by himself, the white employés receiving 1s. 6d., and the natives 3*d*., per carat. Nearly double these amounts are paid for stones found in the mines.

Mr. Gardner Williams, the eminent mining

Mr. Gardner Williams. Lord Randolph Churchill. Captain Williams.

In the Rock Shaft of the De Beers Diamond Mine, at a depth of 900 feet.

engineer who occupies the important post of general manager to the De Beers Company, was kind enough to accompany me all over the mines, and to explain in detail the method of operation. The De Beers and the Kimberley mines are probably the two biggest holes which greedy man has ever dug into the earth, the area of the former at the surface being thirteen acres, with a depth of 450 feet, the area and depth of the latter being even greater. These mines are no longer worked from the surface, but from shafts sunk at some distance from the original holes, and penetrating to the blue ground by transverse drivings at depths varying from 500 to 1200 feet. The blue ground, when extracted, is carried in small iron trucks to the "floors." "These are made by removing the bush and grass from a fairly level piece of ground; the land is then rolled and made as hard and as smooth as possible. These 'floors' are about 600 acres in extent. They are covered to the depth of about a foot with the blue ground, which for a time remains on them without much manipulation. The heat of the sun and moisture soon have a wonderful effect upon it. Large pieces which were as hard as ordinary sandstone when taken from the mine, soon commence to crumble. At this stage of the work, the winning of the diamonds assumes more the nature of farming than of mining; the ground is continually harrowed to assist pulverization by exposing the larger pieces to the action of the sun and rain. The blue ground from Kimberley mine becomes quite well pulverized in three months, while that from De Beers requires double that time. The longer the ground remains exposed, the better it is for washing."[1] The process of exposure being completed, the blue

---

[1] Report, 1890, General Manager, De Beers.

ground is then carried to very large, elaborate, and costly washing machines, in which, by means of the action of running water, the diamonds are separated from the ordinary earth. It may be mentioned that in this process 100 loads of blue ground are concentrated into one load of diamondiferous stuff. Another machine, the "pulsator," then separates this latter stuff, which appears to be a mass of blue and dark pebbles of all shapes, into four different sizes, which then pass on to the assorters. "The assorting is done on tables, first while wet by whitemen, and then dry by natives."[1] The assorters work with a kind of trowel, and their accuracy in detecting and separating the diamond from the eight different kinds of mineral formations which reach them is almost unerring. "The diamond occurs in all shades of colour from deep yellow to blue white, from deep brown to light brown, and in a great variety of colours, green, blue, pink, brown, yellow, orange, pure white, and opaque."[2] The most valuable are the pure white and the deep orange. "The stones vary in size from that of a pin's head upwards; the largest diamond yet found weighed $428\frac{1}{2}$ carats. It was cut and exhibited at the Paris Exhibition, and after cutting weighed $228\frac{1}{2}$ carats. "After assorting, the diamonds are sent daily to the general office under an armed escort and delivered to the valuators in charge of the diamond department. The first operation is to clean the

---

[1] Report, 1890, General Manager, De Beers.
[2] Ibid.

diamonds of any extraneous matter by boiling them in a mixture of nitric and sulphuric acids. When cleaned they are carefully assorted again in respect of size, colour, and purity."[1] The room in the De Beers office where they are then displayed offers a most striking sight. It is lighted by large windows, underneath which runs a broad counter

In the 600 feet level of the De Beers Diamond Mine.

covered with white sheets of paper, on which are laid out innumerable glistening heaps of precious stones of indescribable variety. In this room are concentrated some 60,000 carats, the daily production of the Consolidated Mine being about 5500 carats. "When the diamonds have been valued they are sold in parcels to local buyers, who

[1] Report, 1890, General Manager, De Beers.

represent the leading diamond merchants of Europe. The size of a parcel varies from a few thousand to tens of thousands of carats; in one instance, two years ago, nearly a quarter of a million of carats were sold in one lot to one buyer."[1]

Sorting gravel for diamonds at Kimberley.

The company sustain a considerable loss annually, estimated now at from 10 to 15 per cent., by diamonds being stolen from the mines. To check this loss, extraordinary precautions have been resorted to. The natives are engaged for a period of three months, during which time they are

[1] Report, 1890, General Manager, De Beers.

confined in a compound surrounded by a high wall. On returning from their day's work, they have to strip off all their clothes, which they hang on pegs in a shed. Stark naked, they then proceed to the searching room, where their mouths, their hair, their toes, their armpits, and every portion of their body are subjected to an elaborate examination. White men would never submit to such a process, but the native sustains the indignity with cheerful equanimity, considering only the high wages which he earns. After passing through the searching room, they pass, still in a state of nudity, to their apartments in the compound, where they find blankets in which to wrap themselves for the night. During the evening, the clothes which they have left behind them are carefully and minutely searched, and are restored to their owners in the morning. The precautions which are taken a few days before the natives leave the compound, their engagement being terminated, to recover diamonds which they may have swallowed, are more easily imagined than described. In addition to these arrangements, a law of exceptional rigour punishes illicit diamond buying, known in the slang of South Africa as I.D.B.ism. Under this statute, the ordinary presumption of law in favour of the accused disappears, and an accused person has to prove his innocence in the clearest manner, instead of the accuser having to prove his guilt. Sentences are constantly passed on persons convicted of this offence ranging from five to fifteen years. It must be admitted that this tremendous law is in thorough conformity with South African sentiment, which elevates I.D.B.ism almost to the level, if not above the level, of actual homicide. If a man walking in the streets or in the precincts of Kimberley were to find a diamond and were not immediately to take it to the registrar, restore it to him, and to have the fact of its restoration registered, he would be liable to a punishment of fifteen years' penal servitude. In order to prevent illicit traffic, the quantities of diamonds produced by the mines are reported to the detective department both by the producers and the exporters. All diamonds, except those which pass through illicit channels, are sent to England by registered post, the weekly shipments averaging from 40,000 to 50,000 carats. The greatest outlet for stolen diamonds is through the Transvaal to Natal, where they are shipped by respectable merchants, who turn a deaf ear to any information from the diamond fields to the effect that they are aiding the sale of stolen property.[1] The most ingenious ruses are resorted to by the illicit dealers for conveying the stolen diamonds out of Kimberley. They are considerably assisted by the fact that the boundaries of the Transvaal and of the Free State approach within a few miles of Kimberley, and once across the border they are comparatively safe. Recently, so I was informed, a notorious diamond thief was seen leaving Kimberley on horseback for the Transvaal. Convinced of his iniquitous designs, he was seized

[1] Report, 1890, General Manager, De Beers.

by the police on the border and thoroughly searched. Nothing was found on him, and he was perforce allowed to proceed. No sooner was he well across the border, than he, under the eyes of the detective, deliberately shot and cut open his horse, extracting from its intestines a large parcel of diamonds, which, previous to the journey, had been administered to the unfortunate animal in the form of a ball.

The De Beers Directors manage their immense concern with great liberality. A model village, called Kenilworth, within the precincts of the mines, affords most comfortable and healthy accommodation for several of the European employés. Gardens are attached to cottages, and the planting of eucalyptus, cypress, pine, and oak, as well as a variety of fruit trees, has been carried to a considerable extent. A very excellent club-house has also been built, which includes, besides the mess-room and kitchen, a reading-room, where many of the monthly papers and magazines are kept, together with six hundred volumes from the Kimberley Public Library. There is also a billiard room, with two good tables given by two of the directors. A large recreation ground is in the course of construction. Within the compound where the native labourers are confined is a store where they can procure cheaply all the necessaries of life. Wood and water are supplied free of charge, and a large swimming bath is also provided, but I did not learn if the natives made much use of it. All sick natives are taken care of in a hospital connected with the compound, where medical attendance, nurses, and food are supplied gratuitously by the company. I should not omit to mention that the entire mine above and underground is lighted by electricity. There are ten circuits of electric lamps for De Beers and Kimberley mines. They consist of fifty-two arc lamps of 1000 candle power each, and 691 glow lamps of sixteen and sixty-four candle power each, or a total illuminating power of 63,696 candles. There are, moreover, thirty telephones connecting the different centres of work together, and over eighty electric bells are used for signalling in shafts and on haulages. Such is this marvellous mine, the like of which I doubt whether the world can show. When one considers the enormous capital invested, the elaborate and costly plant, the number of human beings employed, and the object of this unparalleled concentration of effort, curious reflections occur. In all other mining distinctly profitable objects are sought, and purposes are carried out beneficial generally to mankind. This remark would apply to gold mines, to coal mines, to tin, copper, and lead mines; but at the De Beers mine all the wonderful arrangements I have described above are put in force in order to extract from the depths of the ground, solely for the wealthy classes, a tiny crystal to be used for the gratification of female vanity in imitation of a lust for personal adornment essentially barbaric if not altogether savage.

CHAPTER IV

GOLD.

The diamonds of Kimberley—The journey to Johannesburg—Railway extensions—Grass veldt between Kimberley and Vryburg—The cattle farm of the future—"Native Reserve" of the Southern Bechuana—We reach Vryburg—Sir Sidney Shippard entertains us—Coaching with a team of mules—The wayfaring man in the Transvaal—An attractive little town—Gold mines in the neighbourhood—The outlook at Johannesburg—The gold mines—Selfish jealousy of the Boer Government—Astounding inequality of taxation—Bad condition of the roads to Johannesburg—The vicious system of concessions.

I PASSED from the region of diamonds into the region of gold. The "Arabian Nights" character of this statement is justified by the fact that, as the small district of Kimberley produces some 2,000,000*l.* of diamonds annually, so the larger but still not vast district of the Randt produces in the same period some 2,500,000*l.* of gold. This latter production, unlike that of Kimberley, is likely to be doubled and even trebled in course of time. The journey from Kimberley to Johannesburg, covering a distance of 450 miles, deserves some description. The railway is completed as far as Vryburg in Bechuanaland. It is now proposed to carry this railway on a narrower gauge to Mafeking and to a hundred miles north of that place. Should the Tati gold-fields prove as remunerative as well-informed persons believe, the line will soon be extended thither. From Tati to the Victoria Falls of the Zambesi is a short step of about 500 miles, over a country offering few engineering difficulties, and I doubt not that the next generation, before it grows old, will travel to this great river and to its unparalleled cascades with the same ease and comfort as the present generation is able to visit Niagara. The road from Kimberley to Vryburg traverses a succession of plains wide as the eye can range, bounded here and there by low and regular chains of hills. Scarcely a single tree breaks the endless flat of grass veldt.

> Pone me pigris ubi nulla campis
> Arbor æstiva recreatur aura.

The Roman poet must have had Bechuanaland or the Transvaal in his mind when he wrote the lines quoted above, for the two countries perfectly realize his conception. The veldt at the surface in the winter has a somewhat sterile and parched appearance, and is covered with patchy grass dried by the sun to the colour of hay. Far and wide it extends, and the traveller sees no reason why he should ever emerge from its limits. Two causes, however, combine to remove the tedium and monotony of such a landscape. The vastness, the apparent illimitability of the surroundings, elevate rather than oppress the mind, and the genial sunshine, the cloudless sky, the invigorating highland air sustain the spirits at a high level.

Nor must it be supposed that these African plains are in any degree wanting in fertility. The heavy rains of the summer and autumn produce an abundance of juicy grass, on which are raised large herds of cattle and flocks of sheep. Both in Bechuanaland and in the Transvaal the amount of live stock is very considerably less than the area and the soil are capable of sustaining, and it would scarcely be an exaggeration to assert that if, in the course of centuries, all other supplies of meat for the human race should be exhausted, the African veldt could produce sufficient to fill the stomachs of a starving world. Cattle disease, horse sickness, and the sheep scab at present offer formidable obstacles to the rapid multiplication of live stock. It is highly probable that science and sanitary legislation will before long remove or mitigate these scourges of the farmer. Approaching Vryburg, the railroad runs through the " Native Reserve," a large district which has been set aside for exclusive occupation and cultivation by the Southern Bechuana. The soil here is well-watered and of great fertility; abundant crops of mealies (maize) can be easily raised, and many other kinds of grain, potatoes, and various vegetables might be produced in large quantities were the natives given to industry and agriculture. Report, however, speaks but poorly of the Southern Bechuana; idle and insolent in good years, helpless and mendicant in bad, it is doubtful whether he will be long able or permitted to retain his hold upon a territory which is capable of being transformed into one vast garden.

Vryburg is an urban community in its infancy, which may some day grow into a large and thriving town. At present it consists of a number of low buildings of somewhat mean and squalid appearance, constructed of the inevitable corrugated iron, and spreading themselves out irregularly over a considerable extent of ground. Sir Sidney Shippard, the British administrator, who entertained us most kindly and hospitably, occupies a small cottage which many a British mechanic would despise. It is to be hoped, considering the size and importance of our Bechuanaland possessions and the great powers wielded by the administrator, that the British Government, who insist upon retaining their hold over Bechuanaland, will provide its representative with a more suitable and honourable residence. Three hotels offer to the traveller fair but rough accommodation, and in one or more well-filled stores the immigrant or the settler can obtain most of the necessaries of life and such articles as are requisite for the commerce of the interior. From Vryburg branch off many routes north and east and west: westward into Damaraland, or into the great Kalahari desert towards Lake Ngami; northward to Matabeleland and the Zambesi; eastward, which route we ourselves followed, to the Transvaal and Johannesburg. Passenger coaches are for the present confined to this latter route. As we were a large party, with some amount of baggage, we had secured for our private use two coaches, and we passed the four days occupied in the journey to Johannesburg in tolerable comfort. This kind of coaching is an experience which at the present day can only be tried in Africa. The coaches themselves are the most curious productions of human skill. Intended to hold twelve passengers inside, half-a-dozen outside, besides large quantities of heavy baggage, they are constructed of very solid materials hung upon thick springs of leather, and present the most unwieldy lumbering and old world appearance. They are drawn by ten or twelve mules or horses harnessed in pairs. Two men are required to guide the team, the one holding the reins, the other the long whip with which he can severely chastise all but the leading pair. When driving a team of mules the whip is in operation every minute, constant flogging alone inducing these stubborn animals to do their best. At times one of the drivers is compelled to descend from the box and run alongside the team, flogging them all with the greatest heartiness and impartiality. In spite, however, of all this effort and apparent harsh treatment, an average speed of about six miles is all that can be realized. Roads there are none; deeply rutted tracks are followed. When the ruts get too deep for safety the track turns slightly aside, and to such an extent does this sometimes occur that in places the track occupies a width of a quarter of a mile or more. Swinging, bounding, jolting, creaking, straining over this extraordinary route, the coach pursues the uneven tenor of its way, sometimes labouring and plunging like a ship at sea, constantly heeling over at angles at which an upset seems unavoidable; now descending into the deep bed of a "spruit" (creek), now sticking fast in heavy ground, now careering over masses of rocks and stones. The travellers, all shaken up inside like an omelet in a frying-pan, never cease to wonder that the human frame can endure such shaking, or that wood and iron can be so firmly riveted together as to stand such a strain. It may be mentioned that the life of a coach does not exceed two years, that upsets are frequent, and casualties not uncommon. In this latter respect, however, we were fortunate, reaching our destination without the slightest accident or misfortune, our drivers being skilful and the teams on the whole fairly good. Whether South Africa will ever possess proper coach roads is doubtful. Railroads will soon supersede this antiquated method of travelling, and the coach, with its long team of mules, the transport rider with his waggon, and his still longer team of oxen, will soon become things of the past, or be banished to the remotest regions. At present it is possible for any one who cares about the experience to realize most accurately the mode so graphically described by Lord Macaulay, in which our forefathers travelled in England some two centuries ago. Along the road but few human beings are met, human habitations are

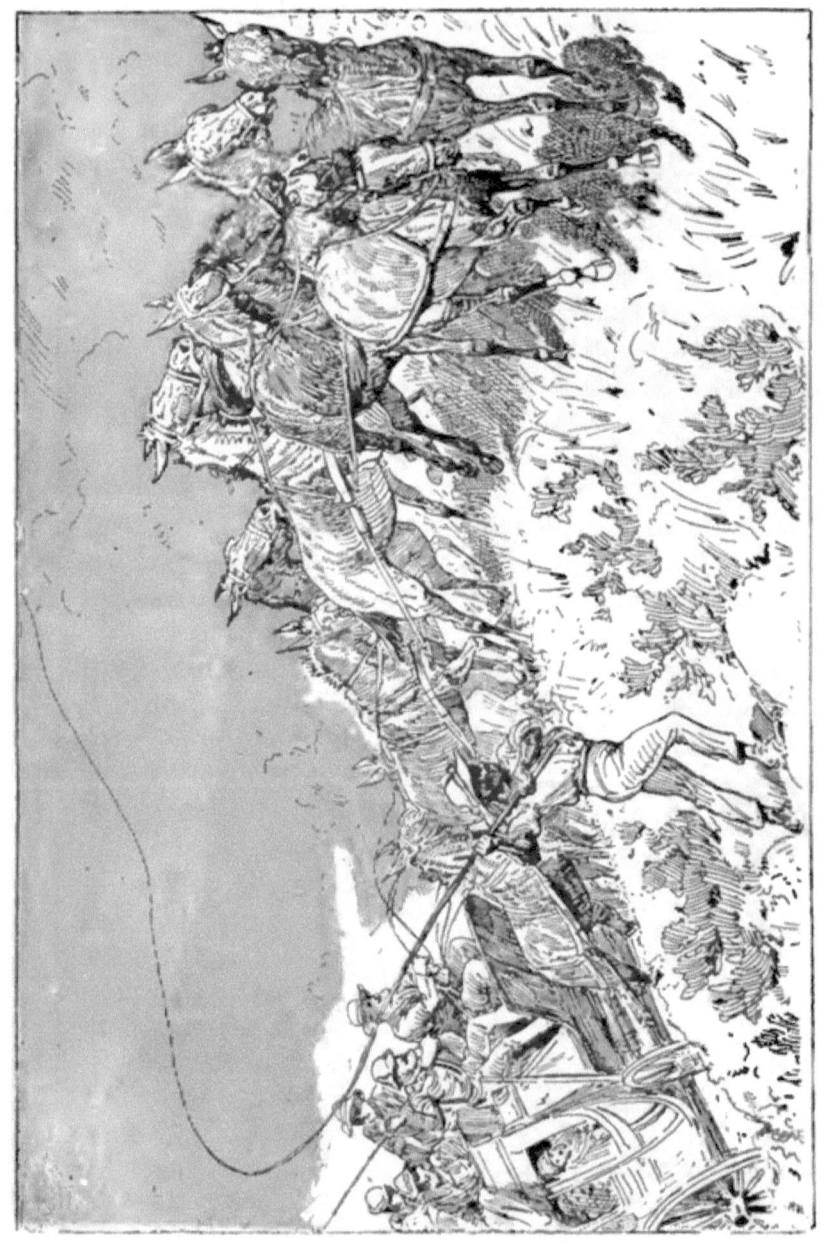

ON THE ROAD FROM JOHANNESBURG TO PRETORIA.—CROSSING A FLOODED RIVER.

scarce and far apart, and little animal life is to be perceived. Birds are fairly numerous; the "koran," the partridge, the plover, the "dikkop," offer to the sportsman occasional shots. Skeletons of horses and of oxen which have succumbed to the labours and privations of the journey abound alongside the track, all either having been picked clean or in the process of being devoured by flocks of vultures. At one place we perceived some scores of these birds surrounding a carcase, so gorged that they took no notice of our approach, although we passed within a few feet of them.

The hotel accommodation in the Transvaal is of the roughest description, the Dutch scarcely appreciating either cleanliness or comfort. It is possible that the sleeping rooms might in some cases be condemned by an English magistrate or inspector. An extraordinary profusion of food awaits the hungry wayfarer, but, alas! it is quantity and not quality which is attained; and it is easy for the man who dines in a Dutch hotel at a table covered with every variety of viand to rise from his repast almost as hungry as he sat down. The following is the menu of dinner which awaited our party on our arrival at Pullen's "Winkel" (store and hotel), where we passed the first night of our journey:—

Oyster soup.
Egg à la soupe.
Saleme (*sic*) curry and rice.
Chicken pie.
Saleme (*sic*) duck and olives.
Roast leg mutton.
Lamb and mint sauce.

Corned mutton.   Saddle mutton
Boiled leg mutton.
Boiled shoulder mutton and caper sauce.
Boiled corned beef.
Boiled ham.
Stuffed roast turkey.
Stuffed roast duck and mushroom.
Roast fowl.
Boiled fowl and oyster sauce.
Potatoes.   Beans.
Boiled currant pudding and wine sauce.
Fruit pie.
Red currant pie and boiled custard.
Tipsey (sic) cake.
Cake à la Méringue.
Custard tart.   Tart.   Queen tart.
Fruit à la Méringue.
Blanc-mange and jam.
Raisins and almonds.

I can truthfully assert that, having done my best to partake of some of these dishes, when dinner was over I would have given a gold mine in Mashonaland for a quarter of an hour at the Amphytrion. On the second day we reached Klerksdorp, and were within the limits of the auriferous portion of the Transvaal. Klerksdorp is in a state of decay, having had but an ephemeral existence. It sprang into life during the gold-mining boom of some four years ago. The ground all round it for a considerable space was hastily pegged off in mining claims, companies were floated with large capital, shares were tossed up to a premium by the promoter, just as a Japanese conjurer with a fan causes bits of paper to ascend in the air, and

THE MARKET-PLACE, JOHANNESBURG.

then came the crash. All was over, and a large pretentious stock exchange, tenanted now only by the dog, the cat, the pig, and the fowl, tells the interesting story of an African golden dream. There are, however, one or two mines in the neighbourhood, which may possibly, with skilful management, yield some profit to the shareholders, where the ore is plentiful, though of low grade. One of these, the Bufflesdorn, about ten miles from Klerksdorp, we were able to make a thorough inspection of. It is being skilfully and economically worked, possesses a reef from three to four feet in thickness, yielding on an average about seven pennyweights to the ton, is not over capitalized, and has much about it which led those who were with me and who were experienced authorities, to suppose that before long it may be a fairly profitable mining venture. Klerksdorp can also boast of a hotel of considerable size, the landlord of which, a German, may be celebrated for his insolence and his rapacity, whose guests are tormented by excessive dirt and discomfort. A short drive of about five hours brought us on the following day to Potchefstrom. This attractive little town lies in the hollow of a great plain, at the foot of some low hills, fairly covered with plantations. The houses are surrounded by luxuriant gardens, the streets are lined with handsome trees. The sight of a tree or of a bit of green in the treeless and parched veldt gives pleasure and relief alike to the mind and the eye. Here a clean and comfortable hotel and an obliging host enabled us to forget the vexations and annoyances of Klerksdorp.

Johannesburg was reached on the evening of the fourth day. We found ourselves at once in a society and amid surroundings widely different from any which had been experienced since landing in Africa. Johannesburg extends for a considerable distance along a ridge of hills 6000 feet above the level of the sea. Around wherever the eye reposes it is arrested by mining shafts, hauling gear, engine houses, and tall chimneys. Johannesburg presents a very English appearance, that of an English manufacturing town minus its noise, smoke, and dirt. The streets are crowded with a busy, bustling, active, keen, intelligent-looking throng. Here are gathered together human beings from every quarter of the globe, the English possessing an immense predominance. The buildings and general architecture of the town attain an excellent standard, style having been consulted and sought after, stone and bricks the materials, corrugated iron being confined to the roofs, solidity, permanence, and progress the general characteristics. The rise of this town has been almost magical, a period of less than five years having been sufficient to effect it; when it is remembered that some twenty millions of capital, mainly English, have been sunk in the mines of the Randt, and that about one and a half million annually is expended on the maintenance and exploitation of the mines, one is confirmed in the belief that there is nothing that money cannot

A STREET IN JOHANNESBURG.

Page 58.

do. The bright days which marked the discovery of the gold mines and the infancy of Johannesburg have passed away. The twenty millions of capital at one time inflated to nearly forty millions, are now reduced to nearer four millions. The London Stock Exchange has become callous and insensible to the attractions of rich reefs, of newly-discovered deep levels, and the inhabitants, many of whom have undergone the bitterest experiences and the strangest vicissitudes, have an aspect to some extent of doubt, nervousness, and anxiety, wondering when the long period of inaction and stagnation, lasting for more than two years, will come to an end, and when their former golden hours will return. I do not think there is any necessity for doubt, care, and anxiety.

Facts speak for themselves, even to a stranger. A gold-field which has been steadily and gradually increasing its output, and which has now attained a monthly production of 60,000 oz., in value some 200,000*l*., must have before it a great future.[1] Recent bad times and the insensibility of the London Money Market have had an admirable effect upon the directors and managers of the gold mines here. They have been compelled, by force of circumstances, to divert their attention from the flotation of new companies and from the pushing up of shares to absurd premiums by making fallacious returns of crushings, and by other dodges familiar to the promoter.

[1] The monthly output for January, 1892, six months after the statement above was recorded, was valued at 84,000 ounces.

They are now concentrating their efforts on the development of their mines, the attainment of deeper levels, the erection of improved machinery, and on economical administration. Many companies have been reconstructed, and others are in process of being reconstructed; the capital of several companies has been ruthlessly cut down. In one case that was brought to my notice, the capital of a company had been reduced from 200,000*l.* to 10,000*l.*, and although this reduction undoubtedly represents a heavy loss in the past, it probably precedes a profitable future. Johannesburg is a town of much promise; the stranger, however, will find occasion for criticism and even censure. The streets are unpaved, and the roadways are as bad or worse than the tracks across the veldt. When the wind is high, as is often the case, the clouds of dust thick and continuous make breathing almost a difficulty, nor is mitigation of this great evil attempted by any kind of pavement, or by the simple and comparatively inexpensive water-cart. The streets at night are unlit, and after sunset total darkness renders locomotion along the bad roads a matter of difficulty and of danger. The obscurity moreover enables footpads and housebreakers to pursue their avocations with considerable impunity, and there has been recently much complaint among the inhabitants in consequence of the increase in this class of crime. The police maintained by the Boer Government, are few in number, nor can I learn that they are in any way distinguished for efficiency. During the week I have been here I have not set eyes on a single policeman, either by night or by day.

There can be no doubt that many of these evils would be promptly remedied if Johannesburg possessed a representative municipality, but the selfish jealousy of the Boer Government obstinately refuses any such concession. There is, indeed, a sanitary board, whose duties are solely confined to matters of sanitation, which is elected by the inhabitants. Two regulations, however, completely neutralize the representative value of this institution. The members of the Board must speak and understand Dutch, and the proceedings and minutes of the Board are recorded in the Dutch language. Johannesburg is essentially an English-speaking town. I imagine that a very small proportion of the inhabitants, practically speaking none of the well-to-do classes, speak Dutch, and thus it happens that those who, from their position and possessions, would be natural and useful members of the Sanitary Board, are totally excluded even from this small share in the government of the town. In the second place the Boer Government nominates the President of the Board from outside the number of those elected. It is not to be supposed that the inhabitants of Johannesburg will long tolerate their condition of absolute servitude in municipal matters. The astounding inequality of taxation between the inhabitants of Johannesburg and those of Pretoria is certain before long to bring about a movement by the former to which the Boer Government will have to yield. Some figures may be useful as showing what our countrymen have to put up with at the hands of the Boers. In Pretoria a Dutchman can purchase a building stand of 26,400 feet square subject to a tax of 1*l*. 10*s*. For the same site an inhabitant of Johannesburg has to pay a tax of 45*l*. The poll-tax in Pretoria is 3*s*. 6*d*.; in Johannesburg it is 18*s*. 6*d*. The Sanitary Board tax in Johannesburg, on a property value 5000*l*., amounts to 41*l*. 13*s*. 4*d*.; in Pretoria there is no such tax. The sanitary fees in Pretoria are 13*l*. 10*s*., in Johannesburg 17*l*. 10*s*. In Pretoria there are no water-rates, in Johannesburg the water-rates on a property of 5000*l*. value amount to 48*l*. The result of these taxation arrangements is that a Dutchman in Pretoria, owning a property value 5000*l*., pays in rates and taxes 15*l*. 3*s*. 6*d*., an Englishman in Johannesburg owning a similar property pays in rates and taxes 153*l*. 1*s*. 10*d*. This system of taxation, so inequitable, so audacious, so impudent, cannot be expected to endure. Pretoria, with a population of 6000, mainly Dutch, is maintained at the expense of Johannesburg, with a population of 15,000, mainly English. The latter, when it has time to give its attention to municipal as well as mining matters, will demand and exact a thorough reform. Two years at the least will probably elapse before Johannesburg, a town whose life and growth depend on the construction of railways, is properly connected with the sea-coast, with other South African towns, or even with all of its own adjoining coal-fields.[1] Millions of tons of machinery, of coal, of provisions, of all necessaries of life, have had to be dragged over hundreds of miles of ground in groaning overladen waggons by exhausted, half-starved oxen. In such a condition of things, one might have thought that the most simple and inexperienced Government could, at least, have maintained decent highway communication. Yet the tracks are the worst in the world, in many places almost impassable at the best period of the year, totally impassable in the wet season. A comparatively small expenditure would suffice to render traffic possible, and even easy. The loss of life among oxen, the wear and tear and damage suffered by and done to wheeled vehicles on account of these awful and even perilous tracks, the loss sustained by a system of transit too dilatory and tedious for description, must be incalculable, and certainly vastly exceeds the amount requisite for the maintenance of proper highways. It is, I believe, the fact that repeated applications have been made to the President for money to be spent on improving or repairing the roads, but all such applications are vain. The President replies that he has no money to spend on such things as roads, that the tracks which are in existence were made by and were good enough for the forefathers of the Boers,

---

[1] Since this was written a railway convention has been concluded between Cape Colony and the Transvaal, under which the railway will be extended to Johannesburg before the close of 1892.

and are therefore more than good enough for the present day. The perverse simplicity of these Boers is inconceivable, but to it there attaches a dark stain. Corruption, it is openly and publicly asserted in the press, in public speeches, and in society, sways violently and malignantly Government circles. The vicious system of concessions abounds. Dynamite, an article of prime necessity in a mining country, has been made the subject of a monopoly, and granted to an individual who, for considerations unknown, is entitled to exclude all other dynamite from the country but his own, and receives a royalty of 12$s.$ 6$d.$ a ton on all his own dynamite which is consumed. To such a pitch has the policy of concession been carried, that I am informed that quite recently an individual applied to the Government for a concession to grant concessions, and that the proposition was gravely and seriously considered, but has not yet been accepted. If this country had been in the hands of the English or the Americans it would probably now be peopled by some millions of Europeans, would be giving forth every variety in inexhaustible quantities of vegetable, animal, and mineral produce, would be intersected by railways and canals—in a word, it might be the most wealthy and prosperous spot upon the face of the earth. But Providence has cursed it with the rule of 50,000 Boers, and for a time, but I expect only for a time, it is destined still to languish.

CHAPTER V.

MINING AND SPORTING.

Account of the Robinson Gold Mine at Johannesburg—The Langlaate Estate—Chlorination at the Ferriera Mine—Dr. Simon—The McArthur-Forrest process—Observations on the gold-fields of Johannesburg—Silver Mines in the Transvaal—Deer preserves—With Dog and Gun in search of Game.

OF all the gold mines round Johannesburg, the Robinson mine is the most remarkable for its size both in respect of area and capital invested, for the high average richness of its ore, for the enterprise and method of its management. This mine was originally bought by a small syndicate for less than 20,000*l*. In 1888, a company was formed to work it, with a capital of 2,700,000*l*. The company possesses a "myn pacht," or mining lease, of about 200 acres, containing some sixty mining claims. Three distinct reefs are being worked at different levels, the main reef leader, the middle reef, and the south reef. The latter has hitherto afforded the richest results. The deepest level now being developed is about 500 feet below the surface, and it has been found by assay, but not yet confirmed by practical crushing, that the ore at this depth maintains its richness. There are about five miles of underground workings, mostly illuminated by the electric light.

F

The ore, which near the surface of the ground is a friable conglomerate, free from pyrites, becomes at the deeper levels hard conglomerate rock, almost impervious to the ordinary drill and hammer worked by manual labour, and highly pyritic. These two qualities have necessitated the installation of American air-drilling machinery of such power as to be capable of drilling a hole four feet deep into the rock in fifteen minutes, which a native would be unable to complete working an

General view of the Robinson Gold Mines at Johannesburg.

entire day. The abundant presence of pyrites compels the chemical treatment of the concentrates and tailings, the stamps alone being unable to extract more than fifty per cent. of the gold. The McArthur-Forrest process, or, in other words, the treatment of the ore by cyanide of potassium, is being tried upon the tailings, and a chlorination plant is being installed for the treatment of the concentrates. It is too soon to pronounce upon the respective merits of these processes, and it is possible that the expense per ton may be greater than would admit of appreciable profit.[1] Here and there in the deeper levels pockets of ore of extraordinary richness are found. I have before me as I write a specimen taken from such a pocket estimated by assay to produce a thousand ounces to the ton. This is probably an exaggerated estimate. Another specimen has been estimated to produce twenty-eight ounces to the ton. The average yield of the ore in the deeper levels will probably be found to be a little under two ounces per ton. The entire gold production of the Robinson mine since the commencement of the year 1889 up to July 1891, a period of a little more than two years, may be stated in round figures at 100,000 tons of ore, realizing 200,000 ounces of gold, in value from six to eight hundred thousand pounds. Upon the enormous capital the directors declared for the year 1889 a dividend of five per cent., and for the year 1890 four per cent. They spent moreover out of earnings on the development of the mine, and on new machinery, an amount equal to these dividends. From October, 1891, when the additional twenty stamps have been erected, making a total of sixty stamps, when the rock-drilling machinery is at work and the chlorination plant set up, the manager expects to get from crushings from 8500 to 9000 ounces of gold per month.[2] There are employed in the

[1] Since the above was written both the processes mentioned have been worked at a good profit.

[2] The returns of the crushings at the Robinson mine for the month of January, 1892, showed a production of nearly 12,000 ounces of gold.

Robinson mine 130 Europeans and about 900 native workmen. The wages paid to Europeans range high; carpenters receive from 5*l*. to 5*l*. 10*s*. a week, skilled mechanics and blacksmiths receive 6*l*. a week. Strange to say, in spite of these high wages, the white workmen are constantly leaving their employment and going off to Mashonaland. The directors find it more and more difficult to obtain skilled labour, and there appears to be, both at this mine and generally all over the Randt, a most promising opening for young English mechanics and miners. The cost of living would probably exceed the cost of living in England, but the high wages, coupled with dwellings rent free, in addition to a magnificent climate, appear to open the road to fortune. The Robinson mine is probably one of the finest gold mines in the world, but it is overburdened with an excessive capital account, which before long it may be found convenient and practicable considerably to reduce. Situated somewhat to the west of the Robinson Mine is the Langlaate Estate. This company, with a capital of 450,000*l*., owns and works an estate held in freehold, not under a mining lease, of considerably larger area than that held by the Robinson Company. The main and south reefs are principally worked, but the average yield does not exceed 15 dwts. to the ton. There is, however, an enormous quantity of this ore in sight, and the excellent management enables a good profit to be realized. A battery of 120 stamps is in process of erection on this mine, which is perhaps the best developed and generally the most attractive of all the mines in the Randt. The Ferreira Mine, adjoining the Robinson, is justly celebrated for its splendid milling plant and machinery, and for its economical and skilful administration. The mine consists of about fifteen claims, yielding, on an average, nearly one ounce to the ton. The concentrates and tailings of this mine, when properly treated, are expected to produce a considerably additional yield. Here has been installed a very perfect assay and smelting plant and laboratory. By the courtesy of the very skilful gentleman in charge of this department, Dr. Simon, I was enabled to follow the beautiful process of the treatment of pyrites by chlorine gas. The pyrites are roasted previously to treatment, becoming extremely friable, losing the sulphur which they contain, freeing the gold, and rendering it accessible to the attractions of chlorine. In the McArthur-Forrest process, or the cyanide of potassium process, the tailings do not require to be roasted, the expense of treatment being thereby considerably reduced, but it is asserted that the McArthur-Forrest process is only available for the treatment of tailings where the gold is free, and that it produces no appreciable results when treating pyritic concentrates.[1] In the simple chlorination process the pyrites having been roasted (sufficiently to make them porous,

[1] This statement, which was made on the authority of Dr. Simon, is altogether denied by the representatives of the McArthur-Forrest process.

but with a slow heat in order not to smelt them), are placed in a vessel upon a filter composed of powdered quartz and glass. The chlorine gas is produced in another vessel by combining manganese and hydrochloric acid. It is then passed through water in order to get rid of the hydrochloric acid, and it is then passed through sulphuric acid in order to get rid of the water which it may have taken up. It finally penetrates through the filter described above, to the pyrites in the condition of pure chlorine gas. In a few hours the chlorine combines with the gold in the pyrites, and becomes chloride of gold. This chloride then treated with sulphate of iron, the gold is immediately precipitated in the shape of a black powder ready for smelting. The process when conducted and viewed in a laboratory is very beautiful and wonderful. Other mines claiming attention, and either now or in process of becoming valuable properties, are the " Simmer and Jack," the " Jumpers," and the " Salisbury," all of which I have had the opportunity of inspecting. Speaking generally about this goldfield, it may be remarked : (1) The ore, when first discovered near the surface, was free-milling ore, easily treated, and yielding in places from two or three up to as much as eight ounces to the ton. Small batteries originally produced striking results, the managers being able to pick and choose those parts of the reef where the ore was richest. Since that time larger stamp batteries have been everywhere erected, the easily-treated rich free-milling ore has become or is becoming rapidly exhausted, and most, if not all, of the mines have now before them an almost inexhaustible quantity of hard conglomerate rock, yielding, when treated as a whole and indiscriminately, a considerably lower average of gold, and to extract the gold from which, with any prospect of fair profit, requires the most ingenious and elaborate appliances and the most skilful and economical administration. (2) In the early days of the Randt gold-field folly and fraud reigned supreme. The directors and managers were, as a rule, conspicuous for their ignorance on all matters of practical mining. The share market was their one and only consideration, the development and proper working of the mine being in many cases absolutely neglected. I was shown the other day the Grahamstown Mine, which, possessing only a claim and a quarter, was palmed off upon the public with a capital of 120,000*l*. This mine, though situated on the main reef, unfortunately struck upon a spot where the reef was intersected by a thick dyke of clay, and it is scarcely an exaggeration to say that hardly an ounce of gold ever has rewarded or will reward the victimized shareholders. (3) But this case is by no means unique. Millions of money have been literally thrown away. Bad machinery badly put up has been badly situated, badly worked. Many of the mines are at a standstill for want of capital, and most of them, so eminent experts assure me, are sadly behindhand with their development in view of the vast plant which has been erected. A healthier tone and spirit now prevail, the work of reorganization, of reducing capital and working expenses, proceeds apace. Unskilful managers and incompetent directors are being got rid of, either by the efforts of shareholders or of far-sighted men, and viewing the extent and nature of the reefs it is safe to predict that the Randt is on the high road to become the greatest gold-field of the world. It should be remembered that in addition to all the difficulties and obstacles which I have described above, and which the gold-fields have had to encounter and overcome, must be reckoned the most stupid, selfish, and incompetent Government which ever afflicted a community or a country. The Transvaal possesses everything which man can desire for comfort, luxury, and general prosperity. An unequalled climate, a soil of exuberant fertility, mines of gold, silver, coal, and iron, all of great richness: the Boers in their stubborn and mulish ignorance have resolved that, so far as in them lies, none of this great wealth shall be taken advantage of and developed. In a country where millions of acres might produce millions of quarters of grain, only comparatively a few hundreds of thousands of acres produce Indian corn. In a country where the storage of water and irrigation works offer little difficulty either to the engineer or to the exchequer, no systematic storage of water is attempted. Yet the presence of water everywhere within a few feet of the surface of the soil, and the long period of winter drought, would seem to render such storage of water and such irrigation works imperative. In a country destitute of trees, but which nevertheless might after a few years' care and industry be covered with forests of various and valuable timber, not an effort at tree planting is made except in the neighbourhood of the gold mines. In a country where for the development of its mineral resources the rapid construction of railways is essential, and where the physical configuration of the ground and other causes marvellously facilitate such construction, the same stubborn ignorance, the same mulish folly before alluded to, has successfully delayed and still delays any such railway construction.

In the foregoing pages I have spoken of the silver mines. These are situated some forty miles to the east of Johannesburg, and are of very recent discovery. The history of them is somewhat remarkable. A company was formed to work them with a capital, I believe, of about a quarter of a million. The affair was probably a fraud, the money was mostly wasted, little was found, nothing was done, and the silver mines of the Transvaal fell into disrepute and disfavour. Some person or persons, however, discovered on the property specimens of ore of singular richness. These being brought to gentlemen possessing experience and capital, were pronounced by them to be good silver ore. A small syndicate was soon formed, shares of the old company were quickly bought up, new capital was expended, the reef has been opened up and developed and ascertained to be of great extent and fair richness. The average yield of the ore has been estimated by assay to be about 30oz. of silver to the ton. In some places, however, it reaches the high average of from 200 oz. to 300 oz. to the ton. It also contains about 30 per cent. of lead. I am informed by experts that the geological formation of these ore deposits is peculiar, the presence in abundance of carbonate of iron and the almost total absence of zinc and of any excess of silica rendering smelting very easy. At present some difficulty in working this ore at a profit may arise from the necessity of having to use for smelting imported coke at the cost of some 15*l*. a ton. In the immediate neighbourhood coal mines are being worked, but it is doubtful whether this coal can be manufactured into coke sufficiently good for smelting purposes. It is known, however, that there exist hard by beds of superior coal, and great hopes are entertained that sufficiently good coke may be produced upon the spot. Silver reefs appear to abound on the properties adjoining that of the Transvaal Silver Mines Company; one or two small syndicates have been formed to acquire and develop these properties, and it is quite possible that the silver mines of the Transvaal may become a larger, a more important, a more valuable industry than even the gold mines of the Randt. I made, in company with some friends, a very interesting and pleasant expedition to these silver mines, and the incidents of the journey lead me to offer a few remarks upon the presence of game and the prospects of sport in the Transvaal. My friend and I, who were naturally not competent to form any practical judgment on mining values, took with us our guns and dogs in order to while away the time during which the engineers and experts would be at work. Not very many years ago these wide and grassy plains abounded with game of almost every description. Persons whose word can be implicitly relied upon have informed me that within the last fifteen years they can remember these plains being covered as far as the eye could reach with countless thousands of wildebeest, blesbok, springbok, and other varieties of the deer and antelope tribes. So desolate and lifeless is the appearance of these plains now that it is difficult to credit the assertion. It happened, however, unfortunately for the sportsman, that not long ago the demand for hides was considerable, and the wise, prudent, and intelligent Boer immediately set to work and slaughtered without discrimination every wild four-footed animal. So reckless and ruthless was the slaughter that these Boer sportsmen (?) never cared to carry home the animals they had slain. Forming themselves into large shooting parties, they shot the beasts down everywhere by scores, by hundreds, and by thousands, leaving the carcases to be devoured by the vultures, and going a few days afterwards to gather up the skins which the vultures had neglected, and which the sun had dried and tanned.

Now the traveller can compass mile after mile of plain without seeing so much as a solitary buck. In a few places, however, attempts are made to resuscitate and preserve the blesbok and the springbok. On an estate of some 80,000 acres belonging to Messrs. Marks and Co., situated on the Vaal river, about forty miles south of Pretoria, there has been raised a herd of a few hundred springbok, which are carefully preserved. On another estate not far off, near Paritj, belonging to Mr. Koetze, some thousands of blesbok are to be found, and are carefully preserved. These two examples show what might be done in the way of preservation of deer if, not only as regards this, but also as regards many other matters, God had only given a glimmer of intelligence to the Boer. For it must be remembered that these animals are fairly profitable to keep, both their meat and their hides being in some demand. Over the whole Transvaal, however, little now remains to the sportsman beyond feathered game. This exists in respectable quantity and variety, but the expanse of plain is so vast that the game is greatly scattered, and the sportsman must often walk far and long before he is rewarded by a shot. On the grassy veldt will be found more than one kind of magnificent crane; the pauw or greater bustard may sometimes be secured with a small-bore rifle; the koran or lesser bustard is more numerous; in the morning and in the evening his discordant call may constantly be heard, and in anything like decent cover he is easily secured with a shot gun. Scattered about the veldt are "pans" of water, surrounded by reeds and rushes, where wild-fowl may often be seen in considerable numbers, but generally difficult to approach; while along the "spruits" and in swampy places, snipe of more than one variety at certain times of the year abound. In the neighbourhood of cultivated grounds, of homesteads, and of gardens coveys of red partridges are frequently met with; and in certain spots quails may be said to swarm at the proper period of the year. The sportsman, however, will have to work very hard and shoot very straight to make up what to an Englishman would appear a respectable bag. My friend and I, shooting for two days in the neighbourhood of the silver mines, obtained the following singularly varied but somewhat scanty bag, nor do I think that the scantiness could be fairly attributed to any excessively unskilful shooting: three snipe, ten quail, six duck, one wild goose, seven partridges, five koran, three plover, four pigeons, one eagle, and five bitterns.

At this time I was enabled by the kindness of Messrs. Marks to make a shooting expedition to the estate before alluded to, which is carefully preserved, and where shooting is seldom allowed. Our party consisted of four guns, and we remained on the estate for four days. The weather was perfect; cool, and even frosty nights, bright and warm days with refreshing breezes. We camped out on the veldt, sleeping in a bell tent. The method of living, though rough and ready, was rendered enjoyable by the presence of a French cook, who skilfully treated us in the evening with the different varieties of game we had secured in the day. In the morning, up before daybreak, mounted on well-trained ponies, we chased the springbok emerging from the cultivated land on to the veldt, galloping as near to them as the fleetness of our horses would permit, generally about 400 yards, dismounting, rapidly firing a snappy chancey shot, then remounting and after them again, getting, perhaps, two or three more shots, and so on until the herd had galloped far away out of range. This method of shooting deer requires much habit and experience, and much good fortune. Only two springbok rewarded our efforts, neither of which, I must confess, fell to my rifle. The chase, however, is in itself exciting, the gallop across the veldt in the cool morning air indescribably exhilarating, and the effect of it is to make breakfast a widely different and far more agreeable meal than one knows it to be at home. After breakfast the shot guns are resorted to, and likely places are hunted over by pointers and setters, or beaten by "boys," after the partridge, the koran, or the quail. In the evening, thoroughly wearied out, the dinner table and the camp fire are found to be real luxuries, and nine o'clock would see us huddled up in our bell tent, and sleeping that slumber which only the satisfied sportsman knows. A description of the bag may be of interest: four duck, fifty partridge, four hares, 250 quail, eight koran, eleven snipe, one dikkop, one wild turkey, one blue crane, and two springbok.

CHAPTER VI.

THE TRANSVAAL BOERS.

> A chance for British enterprise—The capacity of the Transvaal and the incapacity of its rulers—The journey from Johannesburg to Pretoria—Description of Pretoria—The Dutch Parliament—From the Strangers' Gallery—An interview with President Kruger—Parliamentary manners—General Joubert—Report of a case showing the Boer idea of justice—Illtreatment of Natives by the Boers—Shall we surrender Swaziland ?—The Withering Grasp of the Boer.

No English traveller who observes and reflects can leave Johannesburg and not desire that the merits of this town and its many attractions might be made known to and appreciated by the English people. Here almost every description of British enterprise and skill may find a promising opening. The accountant, the young clerk who has received a good commercial education, the skilled mechanic, the farmer, the market gardener, the miner, the agricultural labourer, will all find themselves in demand. A comparatively small capital, ranging from 10*l*. to 50*l*., would probably be found sufficient to start these different descriptions of labour on their road to fortune, relatively, of course, to the employment which they select, and to the education which they have received. Domestic service also offers a most favourable field. Scarcely anything is requisite for success beyond steady and temperate habits, and an industrious and versatile disposition. I have come across more than one young Englishman, who, coming out here at an early age, with hardly a penny at his disposal, finds himself at the age of twenty-five in possession of what may be accurately termed a fortune. Now, probably, is the moment for an emigrant. In two years' time or so a railroad communication between Johannesburg and the coast will have been established, and although by the establishment of such communication, all vested interests from the highest to the lowest, corporate and individual alike, will be enormously increased in value, the labour market will become more fully stocked, and the competition for existence proportionately harder. It can hardly be a matter for doubt that the gold-fields of Johannesburg are destined to attract and support a population which will ultimately dominate and rule the Transvaal. Not only is it certain that there is gold ore practically in sight sufficient to occupy the energies of a mining plant far larger than that which now exists for one or two generations, but the many wants of a mining population where wealth is easily and largely gained, and where luxury and free expenditure become a habit, offer to every variety of commercial enterprise promising prospects. The mere feeding of such a population will be a work of great profit. All over the Transvaal, and especially around Johannesburg, the well-watered and yet easily-drained valleys possess a soil of astonishing fertility, which with ordinary skill and care could produce abundant crops of almost every grain, every vegetable, and every fruit. Whether for housebuilding, for use in mines, or for common firewood, the plantation of trees proposes to a landowner munificent remuneration. Such is the geniality of the climate, such the fertility of the soil, that many kinds of useful and valuable trees are estimated by competent authority to make a growth of no less than ten feet in the course of a year. For the independence of the Transvaal Boers it was truly a most fortunate circumstance that the discovery of the gold-fields succeeded rather than preceded the restoration of Boer independence in 1881. Had Johannesburg, with its present population, its present possessions, and its present prospects, existed at the time of the Transvaal War, it never would have been suffered to pass away from the dominion of the British Government. I adhere to the opinion I expressed in a former letter that the restoration of Dutch independence was necessary if not essential to the peaceful government of the Cape Colony, but viewing the Transvaal as it is, and calculating what it might be if its possessors and rulers were English, one cannot but lament that so splendid a territory should have ceased to be British. The English traveller, according to his disposition, must be sorrowful or indignant when he considers the contrast which is afforded by the capacity of the country and the incapacity of its present rulers. The natural events of the future will probably peacefully retrieve the losses occasioned by the errors of the past. The gold-fields, when connected by railways with the coast, will be crowded in a few years' time with thousands of Englishmen, who will impatiently jerk from their shoulders the government of the Boers. These will be out-numbered, absorbed, or scattered. Already this process is perceptibly going on. All the capital invested in the Transvaal is foreign and under foreign direction. Such is also the case with all industry other than pastoral; I was informed on good authority that more than three-fourths of the land itself is now owned by foreigners. The days of the Transvaal Boers as an independent and distinct nationality in South Africa are numbered; they will pass away un-honoured, unlamented, scarcely even remembered either by the native or by the European settler. Having had given to them great possessions and great opportunities, they will be written of only for their cruelty towards and tyranny over the native races, their fanaticism, their ignorance, and their selfishness; they will be handed down to posterity by tradition as having conferred no single benefit upon any single human being, not even upon themselves, and upon the pages of African history they will leave the shadow, but only a shadow, of a dark reputation and an evil name.

These were the reflections with which I journeyed from Johannesburg to Pretoria. The road traverses a rolling veldt, similar to the other parts of the Transvaal which I have visited. Although a highway of great importance, and crowded with traffic of one kind and another, the Transvaal Government allow this road to be maintained in a condition as bad, if not worse, than that of any other highway communication in the country. Pretoria lies some five-and-thirty miles to the north of Johannesburg. Round Pretoria the veldt becomes more broken, and the eye is pleased at having its range of vision interrupted by low chains of hills, among which are seen pretty dells and valleys and streams, and some appreciable appearance of tree and bush. Fifteen hundred feet less than Johannesburg above the level of the sea, lying in a sheltered hollow, Pretoria, in respect of climate, is far milder and more genial than the high ridges of the gold-fields. The soil of Pretoria is of wonderful fertility for the cultivation of vegetables, flowers, and trees. Nearly every house has its garden, and every garden possesses a plentiful supply of water. The white population numbers under 6000, and it is estimated that some 6000 natives also inhabit the town. It bears all the appearance of a town in its infancy, low straggling cottages and shanty residences adjoining stone and brick buildings of imposing size. The Government buildings, which are approaching completion, erected in a French style of architecture, are distinctly fine and good. They have been built at a cost of about 200,000*l.* They accommodate all the Government offices and the two Chambers of the Volksraad. I was present at one of the sittings of the Dutch Parliament, and observed a combination of pomp and commonplace which was somewhat amusing. The First Chamber (or the House of Commons) transact their business in a lofty, spacious, and well-proportioned hall painted in red and green, the national colours, decorated with the arms of the South African Republic largely displayed, and with a full-length portrait of the present President. The Chairman is seated on a platform which traverses the whole length of the hall. On his right is a seat for the President, and again on the right of the President are seated the members of the Executive, conspicuous among whom was General Joubert. Below on another platform are seated two clerks who read out to the assembly the orders of the day and the contents of bills, memorials, or petitions. To these clerks is also confided the arduous duty of taking down in shorthand the speeches of the members. On the floor of the hall are ranged the members, seated at three long, narrow, parallel tables, slightly curved in the form of a horse-shoe. Some thirty members were present while I was there. Coats and hats were hung up round the wall; a messenger or doorkeeper, in a sort of light brown shooting jacket, heavy walking boots, and a slouch felt hat, strolled about among the members, and represented to my mind the decorous and well-attired officials of our own Parliament. The members speak from their places when called upon by the Chairman. The President is a constant attendant, and takes so free and frequent a part in the debates, that the jealousy of the assembly has been at times aroused, and efforts are from time to time made to restrain the eloquent interference of the Head of the State. I had the advantage of hearing several speeches, and though not understanding the Dutch language, I noticed that the speakers combined fluency with brevity, that their manner was one of ease and of dignity, their gesticulation natural and free. In the Second Chamber I heard the President himself take part in the debate. Three times he spoke with much deliberate composure, but by no means without animation. The two Chambers sit from 9 a.m. to 1 p.m. and from 2 p.m. till 4 p.m. They adjourn, however, for a few minutes every hour, for the purposes of smoking and conversation. During one of the adjournments of the Second Chamber I had the honour of being presented to the President. His Honour is a gentleman of some sixty-five years of age, tall, and rather stout, with a grave, shrewd, but by no means unkindly countenance. At the moment of adjournment he had lit a short pipe, at which he puffed hastily and impetuously. Other members were walking about the Chambers also smoking. Some of these manners the English Parliament might copy with great advantage. His Honour was good enough to express to me the opinion that the Boer trek into Mashonaland, which has been so much talked about, would give rise to no trouble or anxiety whatever, but that, on the other hand, Boer settlers in Mashonaland would be of great advantage and assistance to the Chartered Company. In answer to an inquiry from me his Honour also expressed the opinion that the Matabele would show no hostility against the white settlement in Mashonaland as long as their own proper country was in no way interfered with. The President's manner was extremely gracious and genial, and it was not difficult, after only a few minutes of conversation with him and of observation of him, to understand the great and strong influence which he has acquired and retained over his countrymen. The discussions of the Volksraad in either Chamber are often of extreme simplicity and, indeed, frivolity. The Second Chamber a short time ago discussed at length with much gravity, and at times with some heat and asperity, the question of how its members should be attired. It was at length resolved that a tall white hat, white tie, and black coat should be the prescribed costume. I may add that this Second Chamber has only recently been created. Its members are elected by the same constituencies as those which elect the First Chamber. The Boer Constitution-mongers having brought this political infant into existence were exhausted, and neglected to supply it with powers, rights, or duties. It can neither initiate, nor alter, nor even review legislation. Its consent is not sought for to any law, neither has it any right to discuss any question of expenditure, nor is any information given to it as to expenditure. In its present form it is a mere debating society. In the First Chamber the following incident occurred the other day: Two members, Messrs. Benkes and De Beer, who sit next one another, have the weakness to be exceedingly nervous and shy. Immediately after the afternoon opening, at 2 p.m., Mr. Benkes discovered that some joker had put a dead lizard among his papers. Jumping up he threw the lizard to Mr. De Beer, who loudly exclaimed, "Mr. Chairman, there is a cogolomander here," and ran away. The Chairman: "What is it?" Mr. De Beer: "A lizard, Mr. Chairman." The Chairman: "It won't bite you, it is dead." Mr. De Beer, throwing the reptile at Mr. Benkes, "Take that." The Chairman: "Order, now! let us proceed with the work. Come here, messenger, and take that lizard away." Mr. De Beer then resumed his seat, crying to Mr. Benkes, "You were more afraid than I was." The President, with difficulty sometimes, controls and gets his way with these assemblies. In old days he was accustomed to awe them by threats of his resignation in case they did not agree with him. This method having become weak by over-use, he has hit upon a new device, and quite recently he told the members who where disputing with him that if they did not yield he would reduce their salaries. They were terrified into immediate submission. It may be mentioned that the members of either House receive a salary of 3*l.*per diem while the Houses are in Session. The President receives a salary of 8000*l.* a year. He lives very quietly, never entertains, indeed, he never gives bite or sup to a soul. He is reported to have amassed a large fortune. One of the curiosities of the Boer

Constitution which should be noted is, that during the recess of Parliament the President has power of his own authority to issue proclamations having the force of law, which are, and remain, valid until the meeting of the First Chamber, when they are confirmed or disallowed. This power, which obviously is open to the greatest abuse, has been, it is asserted, much abused.

General Joubert cumulates in himself three distinct offices—that of Commandant of the Army, Minister for Native Affairs, and member of the Executive Government. For these three offices he receives a salary of 3000*l.* a year. Nearly all the offices of Government are occupied by Hollanders. These immigrants—"Uitlanders," as they are called—are disliked by the old Boer and Africander population. They are pure office-seekers, without any sympathy for the Boer, speaking high Dutch—a language "not understanded of the people," and are justly reputed to be as ignorant as they are arrogant, as corrupt as they are stupid. The Boer idea of justice, as between Boer and native, deserves remark. I read the report of a case in which Adriaan E. de Lange, a Government official, belonging to a family much respected in the district, was indicted for having caused the death of a native by violence. It appeared that in November last a Kaffir accused of theft was committed to the care of De Lange, the assistant Field-cornet for the ward of Hoogeveldt, to be lodged in the Rustemburg goal, and that before reaching the latter place the Kaffir died on the following morning, of injuries said to have been inflicted by De Lange. When De Lange was brought before the magistrate he was committed for trial on the charge of culpable homicide, and the magistrate refused to admit him to bail. On hearing of this, the Boer farmers in the neighbourhood assembled in such numbers, and assumed such a menacing attitude, that the magistrate was terrified and allowed De Lange to go out on bail. From medical evidence at the trial it appeared that the Kaffir had been maltreated in a frightful manner, the body being covered with bruises and raw places from top to toe. He had also received internal injuries to the lungs and to the stomach, which were full of blood from ruptured blood vessels; the kidneys were severly inflamed. The external injuries, the district surgeon stated, must have been caused partly by some blunt instrument, such as a "sjambok," and partly by dragging the body along the ground by means of a leather strap which was found attached to the wrist. There was no doubt that death had resulted from the injuries inflicted. The chief witness for the prosecution was Jantje, a native in the employ of a storekeeper, who deposed that De Lange had arrived on the afternoon of November 12th at his master's store with the deceased in charge, and that at De Lange's request Jantje was told by his master to take the deceased to Rustemburg. According to this witness many sores and bruises were visible on the Kaffir on his arrival at the store. Jantje then related in a very graphic manner how De Lange presently overtook him as he was leading the unwilling and weary prisoner along, stating that he was dissatisfied with his slow progress. De Lange, after getting off his horse and thrashing the deceased, got on again, and by means of a long rein fastened to the boy's left wrist, dragged him along the road. This, he asserted, De Lange repeated many times, alternately dismounting to shower blows on deceased with his "sjambok," and mounting again to drag him along the ground by the rein round the wrist. Finally, De Lange, after kicking deceased, and stamping with his foot on his neck, chest, and stomach, left him and rode off to a farmhouse near by to get more assistance, instructing Jantje to go on meanwhile, and if the Kaffir would not walk to drag him if necessary, instructions which Jantje feared to disobey. De Lange presently returning, commenced the same ill-treatment as before, and further seized deceased by the throat, holding him so tightly that the tongue protruded, all but suffocating him. Eventually the poor wretch entirely gave in, and had to be taken to a blacksmith's shop in the neighbourhood, where he was tied up by De Lange, and watch set over him. Jantje's evidence was corroborated by that of his master, by the medical evidence, and by two other witnesses. De Lange then proceeded to Rustemburg, where he spent the night. The following morning he returned to fetch his prisoner, but death was before him, for half an hour previously the hunted wretch had breathed his last. The jury were absent an hour and a half, and on their return announced that they found a verdict of "Not Guilty." The report adds that the accused is a member of the "Gereformeerde," or "Dopper" Church, that all but two of the jury were of the same denomination, and further, that there were relatives of the accused among them. The native, Jantje, whose testimony was so important, quite unshaken under the most searching cross-examination, had been some twenty-six years in the service of his present master, and had always borne an excellent character. For nearly six months before the trial he was detained in gaol as a witness, although the accused was liberated on bail. His master stated that he himself had offered bail to the amount of 250*l.* for Jantje's appearance, but it was refused. The report concludes:—"Among the Boers in the ward, for which De Lange is Field-cornet, feeling ran high, and would, it is thought, have taken very definite shape had the verdict been different. De Lange is most popular with them, for it is felt that he is a man who understands how to deal with a Kaffir. Should he consider that recent occurrences make it becoming on his part to resign his field-cornetcy, they express their determination to re-elect him immediately."[1] Such is Boer justice.

The above case is a typical one, and for that reason, as well as for its shocking details, I have

[1] Local Newspaper, *Standard and Diggers' News*, May 12th, 1891.

quoted it at length. Not that all Boers, or, indeed, many Boers, would be guilty of such inhuman cruelty. It is typical, in the sense that all Boers, whether on the bench or in the jury-box, would act in a precisely similar manner, no matter what the circumstances of the case might be, where Boer interests and Boer life were on one side and native interests and native life on the other. Cases of cruel treatment inflicted by Boers on natives are by no means rare. The Boer does not recognize that the native is in any degree raised above the level of the lower animals. In conversation he describes the native as a "creature." His undying hatred for the English arises mainly from the fact that the English persist in according at least in theory equal rights to the coloured population as are enjoyed by the whites. In the Transvaal no native may travel from one place to another unless he is provided with a pass. In the towns no native may be out at night, unless he is similarly protected. Neither can any native in the Transvaal acquire a title to land. On the other hand, throughout the Transvaal the native enjoys the valuable privilege of being able to purchase and consume in any quantity the most poisonous alcoholic compounds. Taking all these matters into consideration, I can imagine that a British Ministry or a British Parliament may pause and hesitate before handing over to Boer dominion Swaziland and its people. This is the territory which the Boers eagerly covet as giving them additional grazing ground and a fine opening on to the coast. They aver, with what amount of truth I know not, that Swaziland has been more than once promised to them by persons holding official positions under the British Crown. The main cause and object of the recent threatened " trek " into Mashonaland was to put pressure upon the High Commissioner in this matter of Swaziland. "If you will redeem your promise of giving us Swaziland we will drop the trek." Such was always the Boer thought, and such was often the Boer expression from President Kruger downwards. Two circumstances undoubtedly militate in favour of the cession of this territory to the Boers. In the first place the British Commissioner, Sir Francis de Winton, sent out specially by the British Government to inquire into and report upon the condition of things in Swaziland, recommended the cession of the territory to the Boers. In the second place the present arrangement, namely, a joint Government of the country by British and Dutch Commissioners, is unsatisfactory to all parties concerned, cannot be regarded as a permanent one, and could very easily be made unworkable by the Boers themselves.

In spite, however, of these considerations, in view of the utter misgovernment of the Transvaal, of the insolent denial by the Boers of all political and even municipal rights to persons residing in the Transvaal, other than of Dutch birth, strongly impressed with the knowledge of the vicious and cruel sentiments which the Boers entertain towards the native races, I own that it would be with the greatest misgiving and reluctance that I could persuade myself as a member of Parliament to support the surrender to the Boers of the fortunes and destinies of the Swazis; a race, in many respects, of superior quality and promise, one, moreover, which in recent years has fought gallantly side by side with British troops, and has acquired a peculiar title to British protection.

The Boer farmer personifies useless idleness. Occupying a farm of from six thousand to ten thousand acres, he contents himself with raising a herd of a few hundred head of cattle, which are left almost entirely to the care of the natives whom he employs. It may be asserted, generally with truth, that he never plants a tree, never digs a well, never makes a road, never grows a blade of corn. Rough and ready cultivation of the soil for mealies by the natives he to some extent permits, but agriculture and the agriculturist he holds alike in great contempt. He passes his day doing absolutely nothing beyond smoking and drinking coffee. He is perfectly uneducated. With the exception of the Bible, every word of which in its most literal interpretation he believes with fanatical credulity, he never opens a book, he never even reads a newspaper. His simple ignorance is unfathomable, and this in stolid composure he shares with his wife, his sons, his daughters, being proud that his children should grow up as ignorant, as uncultivated, as hopelessly unprogressive as himself. In the winter time he moves with his herd of cattle into the better pastures and milder climate of the low country veldt, and lives as idly and uselessly in his waggon as he does in his farmhouse. The summer sees him returning home, and so on, year after year, generation after generation, the Boer farmer drags out the most degraded and ignoble existence ever experienced by a race with any pretensions to civilization. I have, I must admit, met some persons in Government circles and elsewhere of Boer or Dutch birth who are entirely excluded from the scope of these remarks, whose manners were polite and amiable, who were anxious to show kindness and hospitality, whose conversation was distinguished by original ideas and liberal sentiments. These, however, are but bright exceptions. I speak of the nation of Transvaal Boers as a whole, as I think I have seen it. I turned my back gladly on this people, hastening northwards to lands possessed I hoped of equal wealth, brighter prospects, reserved for more worthy owners entitled to happier destinies; I rejoiced after all that I had seen in the Transvaal, that the country and the people of the Matabele and the Mashona had been rescued in the nick of time, owing to the genius of Mr. Rhodes and the tardy vigour of the British Government, from the withering and mortal grasp of the Boer.

CHAPTER VII.

ON THE ROAD TO MASHONALAND.

The Chartered Company's Station at Fort Tuli—Mining in the Zoutspanburg District—The Progress of the "Spider"—Our first cooking efforts—Hints for sportsmen—Sixty miles without water—A glimpse of Fairyland—We meet Major Sapte and Mr. Victor Morier—Meeting with Captain Laurie at Rhode's Drift—The Bechuanaland Border Police—A "Boer trek"—President Kruger's position—Sir Frederick Carrington and the B.S.A.C. Co's police—Experiment with the new magazine Rifle.

"There is Fort Tuli." Such were the welcome words uttered by Captain Laurie, of the Bechuanaland Border Police, who was riding with me on the morning of Sunday, the 12th of July. I looked up and found that a sudden turn of the road descending to the Tuli River disclosed an eminence about 300 feet high, somewhat resembling in miniature the Hog's Back at Aldershot, surmounted by a group of white tents over which floated in the breeze the British flag. Early in March, 1891, I was in the Westminster Palace Hotel, talking over with Mr. Cecil Rhodes the journey to South Africa which I then contemplated. "There is Fort Tuli," he said, "the first station of the Chartered Company," pointing to a spot on the map before him, and drawing a straight line in pencil from Pretoria to Tuli. He added, "And that is the road you must travel." I own I little thought at the time I should ever get to Tuli, for these long journeys are chancy sort of things, and many difficulties and obstacles often intervene to prevent their accomplishment. Between seven thousand and eight thousand miles I had travelled since leaving London; now only about four hundred miles separated me from Fort Salisbury, in Mashonaland, to which I was bound. A period of eleven weeks was occupied in compassing the greater distance, a further period of six weeks will be taken up in traversing the lesser. Now begins the hard travelling. The country ahead is still in a savage state. No hotels, no stores, no provisions to be bought on the road, beyond mealies, and perhaps here and there milk and eggs and poultry. Everything necessary for the support of the expedition has to be carried along. Before entering upon the composition and the plant of the expedition, a short description of the journey from Pretoria may be of interest. Our party, consisting of Captain Williams, Mr. H. C. Perkins, myself, and a servant, with a fair allowance of baggage, left Pretoria in the early morning of Friday, the 3rd July. We travelled in one of the ordinary coaches of the country, which had been specially retained. Going north the grass veldt is left behind, and the road descends on to the low country bush veldt, passing through hill scenery of much beauty. It is a great relief, getting away from the high grass veldt, with its hopeless expanse, unbroken by tree, bush, or living creature. Now the surroundings

H

captivate the mind. Trees, bushes, glens, glades abound on every side in much variety. Often one may fancy oneself in an English park, or in an English wood. Although winter, numberless evergreen trees, plants, and bushes attract and please the eye. The day passes rapidly travelling through this lovely country. It is midday, apparently, immediately after sunrise, and dusk before one has had time for a brief afternoon slumber. Our first halt was made at the Warm Baths, about fifty miles from Pretoria. Here there are some hot springs, possessing medicinal qualities. The water issues from the ground at a temperature of about 100 degrees Fahrenheit. The baths are of the roughest description, square holes dug in the earth, the sides plastered with mud. However, we found them fairly refreshing after a long and dusty drive. The hotel accommodation is rude but clean, and doubtless the Boer considers the place a perfect Capua for luxury. The next day, passing always through beautiful woodland scenery, brought us at noon to Nylstrom. Here are a Landroost's office, a telegraph station, a hotel and store, all newly erected. The hotel and store were kept by a rascally fellow, who told us we could have no food for two hours, and on being informed that we were provided with supplies of our own, calmly charged us 2*l*. 10*s*. for an hour's use of the common dining-room. In the evening we reached Bads-loop, where we found some clean bedrooms and an excellent store, the property of a young German settler, who was most obliging, amiable, and hospitable, whose charges were moderate. In the morning we journeyed to Eytings, where again we found good accommodation. The place is named after the owner of the hotel. During the two previous days' travelling through the low bush country we had descended to a level of about 3000 feet, but on the third day the road again ascended, and at Eytings reached the altitude of 4500 feet. Starting at daybreak the following morning we passed through Smitsdorp, a flourishing and rising little town, about 10 a.m., and reached Pietersburg at noon, having taken three days and a half over a journey of about 180 miles. Pietersburg is the capital of the extensive district of the Zoutspanburg. Round about, both in the high and low country, many mining enterprises are being carried on. We visited the Mount Marais Mine, four miles from Smitsdorp, and the Palmitsfontein Mine, about six miles from Pietersburg. In the former the ore is of low grade; in the latter it is in places very rich, but uncertain in extent and depth and pockety. It is not probable that either of these mines will greatly reward its owners. Accounts more or less reliable reached us of extremely rich gold findings recently made in the low country Murchison district, about eighty miles to the eastward of Pietersburg, one mine, "The Birthday," having produced some very remarkable quartz specimens. This district is, however, at present very inaccessible, much tormented with fever, horse sickness, and "fly," and some long time will elapse before precise and definite information can be obtained, and some still longer time before any development of the auriferous properties can be made. But the mineral resources of the Transvaal are, indeed, extraordinary. Far and wide all over the country they may be found, and it is difficult to over-estimate the numbers of the population which will at some future day be settled here or the amount of wealth which will be produced.

At Pietersburg our method of travelling had to be changed. The coach was abandoned and "the spider" resorted to. This latter carriage I had purchased from Mr. Nelmapius at Pretoria; it had been specially constructed for travelling in the veldt. The four wheels are high, light in appearance, but of great strength. The seats inside provide for four passengers, and are roomy and comfortable. On the driver's seat three persons can be seated. Over all, projecting well on to the splashboard, is a light canvas covering fitted with windows and with cushioned sides. At night the space between the inside seats is fitted up with the cushions from the front seat, the curtains behind and in front are let down, the windows raised, and a first-rate sleeping apartment and bed are at once secured. These "spiders" are constructed to go over almost any road, and are far more comfortable and less jolting than a two-wheel Cape cart. Eight fine strong mules, an English driver, and a "boy," complete the equipment. Besides the "spider" we had to engage a cart with six mules for our baggage and provisions. From Pietersburg to Tuli there is no hotel and little store ac-

"THE SPIDER."

Page 100.

commodation. This will, probably, be soon provided, as a coach service now passes along the road, either way, twice a week, but the traveller who desires to be reasonably comfortable will do well to rely upon his own feeding, cooking, and sleeping resources, and for a long time to come will find a night passed in the bush very preferable to one passed in the inside of a Transvaal shanty. For twenty or thirty miles round Pietersburg, the high grass veldt without tree or bush is seen, broken here and there by isolated kopjes. We travelled twenty-five miles on leaving Pietersburg, where I may remark we found a thoroughly good hotel, and encamped near a small store kept by a German. Here commenced our first cooking efforts. To collect brushwood and dried dung for the fire, to fill the kettles and boil the water are the first duties; bacon and eggs and bread are the staple of the repast, supplemented by such tinned provisions as may have been brought along. Eggs and bread and milk are very often not obtainable, when biscuits and preserved milk form indifferent substitutes. My party soon became very skilful and expeditious with their kitchen arrangements, and would have breakfast or dinner ready within half-an-hour of outspanning. The weather was perfect, with the exception of one day, when for a few hours we were troubled with a regular Scotch drizzle; the nights were cool, but not cold; the bush country into which we plunged on the second day after leaving Pietersburg, varied and agreeable. Partridges, "pheasants(?)" guinea fowl, and doves can be secured along the route, and form appreciable additions to the daily meals. Any one travelling in this country for pleasure should certainly be accompanied by a couple of well-trained pointers. With these he would often have excellent sport. Long-haired dogs, such as setters, retrievers, spaniels, should not be brought here, as they soon become infested by ticks which cannot be seen or extracted, and which bleed and torture the poor animals, making festering sores, until the dogs fall away in condition, become weak and useless, and often die. A good supply of carbolic oil is essential, as all scratches from thorns, bites, and stings from insects on the hands or face are likely in this country to fester and give trouble unless treated with carbolic oil. Two hours at daybreak and an hour and a half at sunset are the best times for shooting game, which the wild beauty and variety of the bush renders a most exhilarating pursuit. On the second and third days we had to traverse a route totally unprovided with water for a distance of about sixty miles. The abundant vegetation demonstrates that any quantity of water could be found within a few feet of the surface by digging; but wells are looked upon by the Boers as useless luxuries, and unless Nature has provided a "pan" or "spruit," the Boer passes on, at a cost of no matter what amount of suffering to his animals. For twenty-four hours our mules got no water, and consequently reached Jahshaan on the evening of the third day in a very exhausted condition. A night's rest and good water completely restored them. At Jahshaan is a kopje, where there are many guinea fowls. Here also, is a kraal, where relays of mules are kept for the coach service. On the afternoon of the fourth day we arrived at a spot which for beauty of scenery is unrivalled. The abundant presence of palm trees and palm bushes indicated that the tropics had been entered. Many large trees give most grateful shade. The "cream of tartar" tree is a most remarkable growth, in that the circumference of the trunk, from thirty to forty feet, often exceeds the height of the tree itself, and the branches, which are thrown off at the top, are so disproportionately small, when compared with the trunk, as to give to this tree a most grotesque and rather weird appearance. The fruit hangs in pods about the size of a small cocoanut from the branches, and contains a white, creamy substance highly acid to the taste, which the natives aver is a specific in cases of fever. Our camp was situated about 400 yards from the Limpopo. I was strolling along the river bank in the evening with a gun, when I suddenly came upon the most lovely scenery that I ever beheld; I can only describe it as a combination on a large scale of the tropics, Windsor forest, and a fine reach of the Tay or Tweed. If this was situated in Europe it would be the resort of thousands, and would be covered with hotels, villas, and gardens. The setting sun threw on this enchanting spot a light of inconceivable loveliness. It was absolutely fairyland, but the fairies were a few ugly naked Kaffirs. At this place we met Major Sapte, military secretary to his Excellency the High Commissioner, Mr. Victor Morier, and Major Gascoigne, on their way down from Mashonaland.

The former had been sent by the High Commissioner early in May up the Pungwe River with despatches for the Portuguese Governor and for Colonel Pennefather at Fort Salisbury, commanding the Chartered Company's Police. He told me he had left everything quiet and peaceful on the frontier round Massi Kessi, but that it was unlikely that the Pungwe route would be available as an ordinary travelling route for a considerable time, probably not for another year. He added that he had been treated with the greatest politeness and courtesy by all the Portuguese, not only by the officials, but also by the detached groups of Portuguese soldiers who had been encountered on the road. Mr. Victor Morier, who had been present at the skirmish near Massi Kessi, between the Portuguese and the Chartered Company's Police, gave me an interesting account of that incident. It appears that the Portuguese advanced from Massi Kessi to the position held by the police, informed the officer in command of the police that Manicaland was in a state of siege, that all strangers were to be turned out, and demanded that he should evacuate the position. This the officer declined to do, upon which, after a brief interval, the Portuguese, some 400 or 500 strong, natives and Europeans combined, advanced to attack the position, firing the first shots. They were fired upon in return, and after two hours' skirmishing the Portuguese retired with much precipitation and some loss, and so great was their discomfiture that they stayed not in Massi Kessi, some miles distant, where they would have been undisturbed, but evacuated that place also, and, leaving all their stores, scattered away on the route down to the Pungwe. Mr. Victor Morier informed me that the police force of the Chartered Company only numbered thirty-five all told. This place of outspan for the night must also be commemorated by me on account of the wonderful dinner we had that evening. Baked partridges, fried partridge liver, minced koodoo and stewed vegetables, winding up with hot stewed prunes.

The next day we travelled along the Limpopo to Rhodes's Drift, a distance of twenty-five miles. At Morrison's, a small store four miles from the drift, we were fortunate in meeting Captain Laurie, R.A., now in command of the detachment of Bechuanaland Border Police guarding the drift. He conducted us across the Limpopo, and made us most comfortable in his camp for the night. Our cart with our luggage and provisions had sadly broken down, wheel and dissel-boom having been smashed over the rocky parts of the track, and was far behind; without the aid of Captain Laurie we should have passed a night unprovided with food, covering, or shelter. The Limpopo, or Crocodile river, was high for the time of year, the water coming right over the floor of the "spider," and well up on the shoulders of the horse I was riding. At Rhodes's Drift the river is about 120 yards wide, a fine, strong flowing river. The banks are steep, and the crossing was one of some slight anxiety, but, thanks to the assistance of a trooper of the detachment, who stripped off his clothes and led our mules through the water, we effected the passage without loss or damage. Possibly, in a few years' time, there will be a fine iron railway bridge across this river. On the other side of the Limpopo, 500 yards from the

Sir Frederick Carrington and officers of the Bechuanaland Border Police and British South African Company's Police.

river, lay the camp and fort of the Bechuanaland Border Police; to see again the British flag, to feel that at last one was well out of Boerland, was truly pleasant and refreshing. These Bechuanaland Border Police are as fine an irregular cavalry force as could be seen. Composed of men of good education, and in many cases of good family, their training fits them for all kinds of service, enures them to any hardship, makes any difficulty a trifle to them, enables them to confront with resolution any vicissitude of march, bivouac, or combat. They are clothed in a tunic and breeches of dark yellow corduroy, very smart and well fitting, and wear a most picturesque sombrero kind of hat of the same coloured felt, adorned with a red or blue ribbon, according to the particular troop. Black boots, three-quarters up the knee and partly laced over the ankle, complete the attire. They are armed with a Martini-Henry rifle, which is carried with its stock resting in a small leather bucket hanging from the saddle on the right side. Across the shoulder hangs a bandolier, holding fifty rounds of ammunition. A strong, long sword-bayonet is carried on the left side. Haversack, water-bottle, cloak in front, patrol tin in leather case on the saddle, and a thick, warm rug behind, are also added; the whole weighing, with the rider, on an average about sixteen stone. The force numbers about 600 men, divided into five troops. It has been entirely raised and organized by Sir Frederick Carrington, its present commander, and would certainly under him perform the highest services. The men are all well trained in rifle-shooting, many of them being first-rate marksmen. The great smartness of their appearance and demeanour would satisfy even the particular and critical eye of H.R.H. the Duke of Cambridge. At Rhodes's Drift the small detachment quartered there, consisting of Captain **Laurie** and thirty men, had, in a space of three weeks, cleared some acres of bush, sunk a well with timbered sides about thirty or forty feet in depth, erected a circular fort with thick earthworks and timbered walls and wide deep ditch. Underground in the fort was kept the ammunition and other stores. The whole represented an immense amount of hard, incessant labour, and had been effected with an amount of neatness, of ingenious expedient, of fertility of resource that spoke volumes in favour of the skill and science of the officer, of the *esprit de corps* and resolution of the men. What an army we might have in England if only we had no War Office! The B.B.P. are now guarding about 150 miles of the Limpopo in anticipation of the Boer trek. Along this length of river are four or five drifts where detachments are stationed, and where forts have been erected.

Major Goold-Adams described to me the attempt made shortly before by a party of Boers to cross the river. About thirty Boers, the advanced guard of a much larger party, came down to the river, fully armed, intending to cross. They were called to that they would be fired upon by the British force if they advanced, upon which they sent over two or three of their party to parley. They were informed that they could not be allowed to go in unless they signed declarations of their intention to recognize the British flag, and to abide by the laws and regulations of the Chartered Company, and that in no case would any large armed party be allowed to enter. They refused to sign any documents, and in a manner described as most insolent and menacing, declared that they would cross by force. They returned to their party, and once more came down to the edge of the water. A Maxim gun was brought into position by the detachment, and laid on to them, and the officer, Major Goold-Adams, called out that if they proceeded a single step further he would fire. They halted, hesitated, and, prudent counsels prevailing, turned back and rejoined the main body some distance from the river. Here a violent scene is said to have taken place between the leader of the advanced body and the leaders and men of the main body. The latter were reproached by the former for cowardice and desertion of him. The quarrel terminated by the small and violent group abandoning the enterprise and disbanding. The other and larger body, with whom was Colonel Ferreira and a certain Malan, a son-in-law of General Joubert, then marched to another drift, where they encountered the same officer, and where a similar, but much less stormy, scene took place. Colonel Ferreira crossed over by himself and was immediately arrested under orders received from the High Commissioner, and sent to Fort Tuli. After a few days' detention he was liberated and allowed to proceed up country, having signed all the necessary documents. The other Boers, finding the British in force, determined to resist their passage, retired and immediately dispersed. Thus, happily and fortunately, ended the celebrated "Boer trek." At one moment an encounter, with certain bloodshed and loss of life, was very near, but the firm determination of Major Goold-Adams and his men, the adequate preparations made beforehand by Sir Frederick Carrington and the High Commissioner, averted what would have been a great calamity. "The Boer trek" promised at one time to be a very formidable business. The Boer leaders, more or less encouraged by General Joubert, who were carrying on intrigues with the Portuguese on one hand, and the Matabele on the other, undoubtedly saw their way to a successful incursion into what they regard as "a promised land specially reserved for them by God." Fortunately President Kruger never hesitated; from the first he exerted against the "trek" all his great authority, he kept from it all actual sympathy or effectual support among the mass of the Boers, and his telegram of April last to the High Commissioner to the effect that he had damped the trek was, even at that time, strictly accurate. It is quite possible that in taking this action he has overstrained his influence and imperilled his popularity. Unless he succeeds in obtaining Swaziland for his people this will surely be found to be the case. But these things cannot be determined until 1893, when the next Presidental election takes place.

At Fort Tuli our party was most hospitably received and entertained by Sir Frederick Carrington, Captain Leonard (in command of the post), Major Tye, the civil magistrate, and by the officers of the B.B.P. In the fort are quartered from eighty to ninety men of the B.B.P. and of the British South African Chartered Company's Police

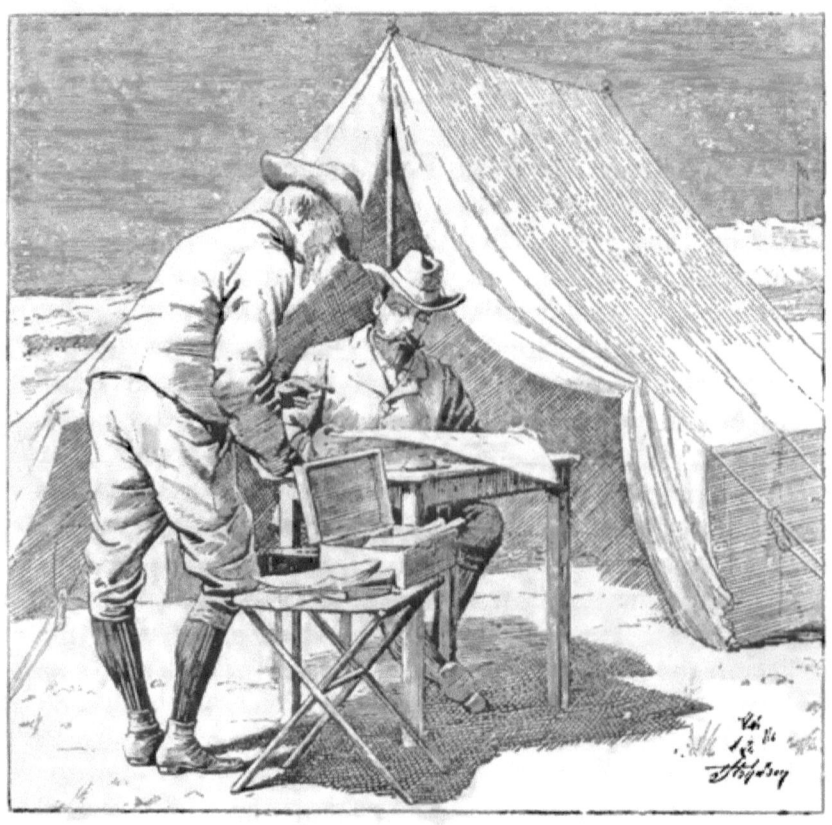

Lord Randolph discussing his route with Sir F. Carrington at Fort Tuli.

(B.S.A.C.P.). This latter force greatly resembles the B.B.P., on the model of which it was formed. The composition of the rank and file of the

B.S.A.C.P. is fairly indicated by the following authentic anecdote:—A new officer had joined and was riding along in front of his men. A trooper riding behind was overheard to remark to another, "I say, Bill, I don't think much of this new fellow. I don't remember having ever met him in White's or Boodle's." There are in the force serving as troopers two sons of British peers, and many men of birth and good family. Some come out to see life and adventure, and make a fortune; others, and not a few, to make a living, and if possible regain a lost fortune. Fort Tuli is a strong position against any artillery which is likely to be brought against it for many years to come. It is armed with a Maxim and with a Gatling gun. Sir Frederick Carrington allowed me to see these guns at practice. The range was 1600 yards, the target some small bushes growing on the sandy bed of the river, which for some distance is effectually commanded by the fort. The Maxim appeared to be remarkable for its precision, the Gatling for the extent of ground swept by its projectiles. The effect of the fire of either was very striking, and I would imagine terrifying to any finding themselves within the range of these ingenious little monsters of destruction. Here I had a good opportunity of ascertaining the opinion of trained marksmen upon the new magazine rifle now being supplied to the British army. The Secretary of State for War had given me one of these rifles, Mark I., to take along with me and try. It was now produced and examined by the officers with

An Experiment with the Magazine Rifle.

much interest. A fine experiment was made with it, one which could not have been carried out in England without the intervention of the S.P.C.A. A slaughter ox was tethered on the sand of the river 1500 yards distant and about 300 feet below the bastion from which the rifle was fired. Captain Capper, renowned in the B.B.P. for his skill as a rifle shot, fired at this distant and certainly not large object. All his shots were observed through the telescope to go very close to the ox. The afternoon was clear, there was no wind. At the twentieth shot the animal fell like a mass, and remained perfectly motionless. We mounted our horses and rode out to examine the carcass. The bullet, which had slain the ox so instantaneously, had entered the nape of the neck rather high behind the ear, passing downwards, severing the spinal cord, and emerging lower down the neck nearer the shoulder on the other side. We observed that the animal had also been struck by another bullet, which had penetrated the middle of his side, passed across the body somewhat upwards, emerging just under the hump on the other side, injuring the intestines and other vital parts. This small bullet had produced no apparent immediate effect on the animal, who had during the firing been under the observation of the strongest telescopes, and was not observed to start or even to make a movement till the last bullet struck him. I asked Captain Capper what he thought of the weapon for accuracy: he told me he thought he would have made more accurate practice with the Martini-Henry, but this he attributed to the method of sighting adopted for the magazine rifle, which he strongly condemned. I think he rather liked the rifle on the whole. On the other hand, I fancy I may state that the balance of opinion was not favourable to the weapon. All pronounced it very complicated; all doubted whether it was a practical weapon for a common soldier. The method of half-cocking the arm, the arrangement and spring of the magazine, the short cleaning rod, the poor and weak bayonet, received nothing but condemnation. One officer, perfectly entitled to give an opinion, said he would like the rifle without the magazine. I thought this the most damaging opinion I had yet heard given. I am confident that all were unanimous that if they had to fight for their lives they would choose the Martini-Henry in preference to the new magazine, but this judgment, definite and unqualified as it was, is perhaps discounted by the notorious human prejudice in favour of what is accustomed to and against change and novelty. This discount, moreover, is strongly supported by the equally notorious fact that at the time of the introduction of the Martini-Henry into the service, high military and high expert opinion leaned heavily towards a preference for the ancient Snider. Again, on the other hand, the defects of the new rifle are great and glaring even to eyes by no means expert, and to minds not trained in mechanics. The uses it will be subjected to, the hands in which it will be placed, cannot have received real practical atten-

EXPERT OPINION ON THE NEW ARM.

tion. Impossible perfection has been sought after irrespective of matter-of-fact practical commonplace considerations. Personally I venture to sum up the question by the remark that it is one of extreme difficulty; that if I were Secretary of State for War, viewing the expenditure to be incurred, the great national disasters certain to follow on an error of decision, the serious and to a great extent successful manner in which the new rifle has been impugned, no human power that I am aware of would induce me to assume the responsibility of imposing this magazine rifle on the army. The Small Arms Committee and other highly-paid expert and inexpert officials with which our country is blessed or oppressed have taken five years to decide upon a weapon. After such an extravagant consumption of time, a few months more would be of little account. A review of all the circumstances of the case by fresh and equally well-informed, but by more impartial and less personally interested judges, would probably allay public anxiety, increase military confidence, and certainly relieve the load of responsibility which must attach to any minister or ministry who make the final decision. Nor can it be said that there is any great hurry. A good magazine is probably a better weapon than a Martini-Henry, but the difference is minute and insignificant compared with the difference between a known and tried Martini-Henry and a bad magazine.

CHAPTER VIII.

THE EXPEDITION: ITS COMPOSITION AND EQUIPMENT.

Major Giles—A fine collection of giants—Our rifles and guns—Warning and advice to future travellers—Composition of the Expedition—Major Giles's trek from Vryburg to Tuli—The horse sickness in Africa—A camp fire concert at Fort Tuli.

At Tuli I had the pleasure of joining my waggons and of seeing again the other friends who accompanied me to Mashonaland, whom I had taken leave of at Cape Town more than six weeks previously. They had been doing all the real hard, rough work of the journey, and making a long, tedious, and, from some points of view, an anxious trek. Major Giles, an ex-Artillery officer of many years' South African service and experience, had undertaken the superintendence and general management of the Expedition: a heavy and complicated business, as will be seen when the composition of the Expedition is gone into in detail, in which he had been most efficiently assisted by Mr. Edgell, who had seen much wild life in the Rocky Mountains and in cattle ranches, and by Mr. McKay, who last year formed one of the Pioneer force despatched into Mashonaland.

EXECUTIVE OFFICERS OF THE EXPEDITION.

I may mention that Major Giles stands 6ft. 4in., Mr. Edgell 6ft. 4½in., Mr. McKay 6ft., Messrs. Mockell and Myburgh, the conductors, 6ft. 5in. each: a fine collection of giants. The organizing and equipping of an African expedition is an elaborate and costly business, and a detailed ac-

The long and the short of it.

count of the work may be of value to those at home who may be contemplating, or who may undertake a similar journey. In London a large outlay had been made. Tents, all camp equipment, cooking appliances and utensils had been supplied by a well-known London outfitter. The following rifles and guns had been bought of a London firm :—1. A double-barrel ·577 B. L. Henry rifle. 2. A single-barrel ditto. 3. A single ·450 B. L. Henry rifle. 4. A pair of No. 12 breech-loading shot guns with rebounding locks. 5. Six Winchester repeating rifles, new pattern, ·450 bore, carrying four cartridges in the magazine. I also had from Messrs. Fraser, an Edinburgh firm, a ·500 bore B. L. double-barrel rifle. With this rifle I did all my shooting, and found it to be a most perfect, accurate and beautifully sighted arm. In addition to this armament there were purchased at Kimberley two pairs of No. 12 shot guns, made by Greener, four ordinary Martini-Henry rifles, and two sporting rifles. We had with us about 10,000 rounds of ammunition. A London house had furnished a great variety of provisions, tinned meats, pressed vegetables, fruit, bacon, ham, tea, coffee. Saddlery, horse clothing, and halters were purchased in London, as also medicines, etc. I would venture to give a word of warning and advice to those who start on a South African journey, and who have to purchase material at home. I foolishly imagined that if I resorted to West-end tradesmen in London, though I would have to pay considerably higher prices, at least I would obtain the best articles turned out and packed in the best possible manner. But in this I was disappointed from not having personally seen after everything, down to the smallest details. For instance, three bell tents were supplied, all of old and each of different patterns, with poles too long, causing very great inconvenience when unpacked and brought into use. All the packing cases were of such weak and flimsy material that after being opened they became useless. More than that, the packing of the articles was so defective that many things were broken, especially an elaborate stove, and lamps of more than one kind. An expensive canteen, on being opened, was found to be defective in many articles. I could cite other instances of carelessness and neglect, which ought to be most carefully guarded against, for in a country such as this defects in the original equipment cannot be made good, will always produce vexation and inconvenience, may sometimes be attended with consequences still more serious. At Kimberley servants and grooms were engaged, waggons, oxen, mules, horses purchased. Here again I would advise the traveller who has to make purchases at Kimberley to personally inspect and examine every article ordered and to see to the packing of it. One large wholesale house to whom I had special letters of recommendation, supplied us with many shocking bad articles of the most shoddy description. Also some essential parts of the mining equipment which had been ordered were found on arrival here not to have been sent. The state of the expedition as I found it on arrival here was as follows:—In addition to those gentlemen I have already mentioned, it had been joined by Captain the Honourable Charles Coventry, of the B.B.P., who had obtained three months' leave. Also I had been fortunate in securing the services of Mr. Hans Lee, a well-known and most successful hunter, through whom I hoped to obtain some good big-game shooting. The remainder of the *personnel* was as follows :—3 white servants, 2 " Caye boys," 4 grooms, two cooks, with 2 native boys to assist, 2 donkey herds, 14 native drivers and leaders. The live stock consisted of 103 oxen, one slaughter cow, 13 riding horses, 18 mules, 1 mare to run with the mules, 14 donkeys, 11 dogs, mostly curs. The vehicles were 1 " spider " carriage, 1 large mule waggon on springs drawn by 12 mules, 4 half-tent waggons, drawn by 18 oxen each, 1 buck or uncovered waggon, also drawn by 18 oxen, the Scotch cart, a covered waggon on two wheels, drawn by 8 oxen. This quantity of wheeled vehicles and cattle and mules had to draw about 21,000 lbs. of meal, mealies, potatoes, onions, and various other provisions, 2000 lbs. of ammunition, 1500 lbs. of trading goods, 2500 lbs. of mining tools and plant, 8000 lbs. of baggage, 5000 lbs. of camp equipment, furniture, and miscellaneous articles, 3000 lbs. of corn and forage for horses, and about 1500 lbs. of saddlery and stable equipment, making a total, with allowances for other necessary weights, of upwards of 40,000 lbs., or, according to local measurement, some twenty tons weight of freight. The enumeration of the above will be sufficient to indicate the amount of thought, care, and trouble requisite for the conveyance of such a troop and such a quantity of stores across such a country as South Africa, with its hopeless roads, its swamps, its rocky places, fevers, and sicknesses, without incurring accident, damage, or loss. The trek from Vryburg to Tuli, a distance of 550 miles, was accomplished in a period of fifty-four days, only thirty-five days of which were occupied in actual

Camp life at Tuli.   Branding cattle.

treking, thus covering the distance at the rate of about 16½ miles per travelling day. This trek was, moreover, accomplished without the loss of a single ox, with the loss only of two mules, one from sickness, one from accident, and with the temporary loss of seven donkeys, five of which have been recovered. No case of sickness has occurred among the men of the expedition, either white or coloured. Major Giles was fortunate with the horses, all of which were brought as far as Tuli in even better condition than they were in when they were originally bought.

The horse sickness in South Africa causes such heavy loss that I am tempted to dwell on this subject. Most authorities are of opinion that it is useless to purchase horses for African journeys, unless they are what is termed "salted," that is, have had and have recovered from the sickness. Such horses, however, are, for the most part, sorry, wretched steeds, without spirit, with very inferior strength. They by no means enjoy perfect immunity from further attacks of sickness. Large prices, moreover, ranging from 50*l*. upwards, are asked for them. Major Giles resolved to ascertain whether by great and constant care he could not, at least at this season of the year, preserve his horses from the sickness. He had to encounter a great deal of derision from persons of all sorts of experience, who freely prophesied he would not bring a horse alive to Tuli. Mr. McKay, who undertook the charge of the horses, gave the following details of his management. First, the horses are never watered before 11 a.m. or after 3 p.m. This precaution is adopted against the evils occasioned by the morning and evening dews, at times and in certain places very heavy. Secondly, when outspanned, the horses are covered with a horse-rug, buckling over the chest, and with a blanket rug, doubled, coming well back over the loins. At sundown the horses are fed in nosebags, the bottoms of which have been carefully tarred. Three times a week each horse has its nostrils slightly tarred inside, once a week a tonic dose is administered to each, composed of about two wine-glasses of gin, with enough quinine to cover a shilling, well piled up, mixed with the gin. Further, in places with an evil reputation for horse sickness, the horses were never allowed to go to the river or other water; buckets of water were brought to the camp and allowed to stand for an hour or more in the sun, and then slightly chilled by mixing warm water. The great and principal precaution is that some trustworthy person should daily see that the grooms carry out these regulations conscientiously. A few minutes' neglect destroys the effect of all the care of days and weeks. I admit that many persons assert that all precautions against horse sickness are unavailing, and that we were favoured by singular luck which could not be expected to follow us long.[1] Yet the treatment described above is strictly in accordance with common-sense and with elementary sanitary science, and is surely worth a careful trial in view of the immense value of horses to the traveller in South Africa. On one day while at Tuli all the oxen were brought in for inspection and appeared to be of fine quality and in first-class condition. Certainly it would not have been thought that the respective spans

[1] This opinion turned out to be correct.

had been engaged during six weeks in drawing waggon-loads of about 7000 lbs. apiece over a distance of 550 miles along Bechuanaland roads.

The camp was, by special permission of the commanding officer, pitched on the north bank of the river, on a space which had been cleared for a cricket ground. All around is the bush veldt, where at some distance from the camp the animals find good grazing. Here, at an altitude of only 1850 feet, the weather is found to be much warmer in the daytime than in the high uplands of the Transvaal, nor is there any frost at night. At this time of the year the situation is fairly healthy, and there is no fever among the troops. During the rainy season the troops suffered considerably from fever and dysentery, the horse sickness ravaged the mounts, some 80 per cent. of horses having been lost. It is said that the Chartered Company will give up this station, which is to be taken over by the Bechuanaland Border Police.

Before our departure the military force entertained the expedition at a camp fire concert. A colossal and Plutonic bonfire threw a wild and glaring light upon the surrounding scenery and upon the groups of men and natives in many-coloured and motley attire. The attendance must have numbered over a hundred. Many excellent songs were sung, one recitation bearing on Sir Charles Warren's Bechuanaland exploits achieved a great success. A single verse will indicate the spirit of the poem and the reputation of the officer:—

A CAMP FIRE CONCERT AT FORT TULI.

So you see there was no fighting, on that glorious campaign,
For not a man was wounded, not a warrior was slain;
And the doctors had an easy time, as doctors always will,
Campaigning with a General who goes fighting with a quill.

It was after eleven before the programme was completed, officers and men taking equal parts in the performance. The men were in the highest spirits, the officer being obviously extremely popular. At the close Sir Frederick Carrington addressed them in a stirring speech, and was enthusiastically cheered. Truly an impressive scene. Here, some thousands of miles away from England, in a country inhabited by a numerous tribe of savages of noted ferocity, not a hundred miles from the kraal of the great Lobengula, was a tiny group of men holding their own, maintaining their authority partly by their own reputation for efficiency, partly because they represented the might and prestige of the Empire; never dreaming for a moment that a shadow even of danger could approach them, never doubting their ability to dissipate any danger should it arise. This is the group of military force which holds for England a portion of South Africa, from Kimberley to Fort Salisbury, comprising a territory as large as Germany and France, replete with elements of a hostile and dangerous nature. May good fortune ever attend and reward them.

CHAPTER IX.

THROUGH BECHUANALAND.

Cold nights in camp—The horse sickness—Visit from Kaffir women to our Mariko River camp—Outspan on the banks of the Crocodile River—We cross the Mahalopsie River—Dr. Saur and Mr. Williams—Camp at Silika—Arrival at the Lotsani River—The luxury of a shave—The Suchi River—Headquarters of the Bechuanaland Police at Matlaputta—The Macloutsie River—I lose myself near the Semalali River while in quest of game—Catching up the waggons.

From the Journal kept by SURGEON HUGH RAYNER.

RAMATLABANA, *Sunday, May* 31*st.*—We are seventeen miles north of Mafeking, out of British territory, but in the British Protectorate. Nights are very cold. As soon as the sun goes down the temperature changes, and after midnight the cold is intense and continues till sunrise, when it gradually becomes warmer. There is, however, always a cool breeze during the day, so that the heat of the sun is considerably tempered. If by chance the sun becomes obscured by clouds a feeling of cold is at once experienced. As a specimen of the night cold, I slept last night in a camp bed with a cork mattress and three blankets. I was in a rough flannel sleeping bag and covered with two camel's hair blankets and a sheepskin kaross. Yet my feet never became warm, and were quite cold on waking this morning. Mr. Sinclair came across to our camp this morning, and kindly offered to take us out for some duck-shooting, so we all made a start on horseback to some "vleys" some few miles away. We came across a flight of seven duck, all of which we killed, after following them backwards and forwards from "vley" to "vley."

Wednesday, June 3rd.—At 1 a.m. we inspanned, and had not proceeded more than a couple of miles before one of the waggons stuck in the mud in a drift. The night was very dark, the moon being in its last quarter. Then two others stuck. Eventually one of them—the meal waggon—had to be unloaded and the others double-spanned before they could be extricated. We did not start again till just before daylight—a hard night for every one except myself. I had a comfortable night's rest, and being very tired from my exertions of the previous day, slept on quietly in my "Kartel,"[1] all unconscious of what was going on outside. We had intended to reach Sandpits by daylight, which is the nearest water (supposed), but luckily we came upon a "vley" where there is seldom water, about nine o'clock, so we outspanned there. . . . At 3.30 p.m. we inspanned, and at 5.30 arrived at Sandpits. On the way one of the mules was attacked by the dreaded "horse" sickness, and was dead in three hours. This sickness is well known in South Africa. It attacks horses and mules suddenly, but donkeys are exempt. An animal is quite well up to a

[1] Large waggon slung mattress.

certain time, in fact, it may be in rather better fettle than usual, when suddenly it appears unwell. It ceases to work and becomes very tottery. In a few minutes it is noticed to be breathing hard, and its nostrils working are evidences of great distress. Almost at the same time a discharge of mucus appears at the nostrils, which presently becomes very profuse. The distress increases, and in a few hours the animal, becoming weaker and weaker, and more and more distressed in its breathing, falls down and dies. Post-mortem shows general congestion of the internal organs, especially of the lungs. All kinds of remedies have been tried, and have failed. In the case of our mule, half a bottle of gin and a large tablespoonful of quinine were at once administered, and this seemed to revive it for a time; but soon the weakness came on again, and the animal died.

June 11th.—Sequana is about fifteen miles from Maripi, our last halting place. It is on the banks of the River Mariko, which provides good water. We are outspanned about 200 yards from the river by the road side. This afternoon a lot of Kaffir women came round with milk, pumpkins, etc., for barter. They were a very good-natured looking lot. I happened, at the time, to be reading the special number of *South Africa*, which contains many excellent pictures of this part of the world and of the various tribes. I showed them to the women, and they recognized several specimen portraits. The first was a picture of Matabele women, correct in detail, because it was copied from a photograph by Surgeon-Major Melladew. One woman immediately recognized it, and clapped her hands, calling out, "Ha, ha! Matabele, Matabele!" Then came some pictures of soldiers, which they also recognized, and with which they were equally pleased. A pleasant half-hour was thus spent. A Kaffir man sold me his hat for 6d., which I took a fancy to, and which was simply the skin of a very pretty little red and black bird, tied jauntily on the left side of his head with a piece of string. Then he went away, but soon returned with another "hat" on. This, however, was not nearly so pretty, and I made no offers. I have no doubt that had I bought it, he could have appeared in any number of "hats" in succession. I also bought from him a jackal's tail (used for brushing flies away) for 6d., and my Kaffir friend went away delighted with his bargains.

Tuesday, June 16*th*.—Inspanned 2 a.m. Arrived at Palla Camp—seventeen miles from No. 4 Post Station. Our outspan is on the banks of the Crocodile River, about fifty yards distant. The actual camp of the Bechuanaland Police is about four miles further on, where there is also a telegraph station. A small detachment of the police are stationed there. There are two stores, one of which is within half a mile of our outspan. I was told that there was a good deal of fever about here, contracted at the close of the late rainy season, but there is always more or less fever along the banks of the Crocodile. The country up to now has been getting gradually more wooded since Vryburg, and the trees getting gradually larger. The thorns all the way have been very troublesome, especially the well-known "wait-a-bit" thorn. . . . We found several waggons outspanned here, Mr. Winslow and party among them. Went out four hours with a rifle in the morning, but saw nothing. In the afternoon Mr. Winslow came and showed us the way to a large "vley" about two or three miles away, where there were numbers of duck and teal. Here we shot about a dozen birds, but Giles was the only one who managed to bring his bird to bag, a very large duck; in fact, almost as big as a goose, with a very broad span of wings. All the other birds fell into the "vley," and it was too deep to wade for them. Also crocodiles were said to live there sometimes. Darkness brought an end to our afternoon sport, so we returned to camp, feeling we had rather wasted our cartridges, and killed birds for no reason. Mr. Winslow and three of his party

Fording a river.

came to supper, and we had a "smoking concert" over a roaring camp fire. . . . I was called out to see a Kaffir "boy" who had been shot in the leg by a man, "X.," for mutiny. The man had pulled out a knife, and meant mischief. He was well peppered in the calf of one leg, and I don't think he will be able to sit down with ease for a few days. However, he was not seriously hurt, as, of course, "X." took good care not to shoot till he was, so to speak, at a safe distance.

Friday, June 19th.—Our outspan is about fifteen miles from our last halting-place, and we are still on the banks of the river. There is a post-changing station close by, and from here bullocks are used for the post-cart instead of mules. This is on account of the dreaded horse sickness. . . . We crossed the Mahalopsie River this afternoon. There was no water in it, simply a dry, sandy bottom. Mackay and I were walking across together when he drew my attention to two depressions in the sand in the middle of the river-bed. These were about a couple of yards in diameter and a couple of feet deep. "See," said Mackay, "some one has been digging for water here. I'll bet there is water about a foot deeper. I'll show you." He then commenced digging vigorously with his hands and shovelling the sand out. Sure enough, about a foot deeper, water flowed into the hole. "That's worth knowing," said he, and we proceeded on our way.

Tuesday, June 23rd.—Inspanned at 1 a.m., and

trekked out twelve miles, making with last night's trek about twenty miles from our last outspan. Arrived at sunrise at the Wegdraai (pronounced Vechdri). This means in Dutch the parting of the river from the road. Giles tells me last night's trek was a very good performance, but of course our oxen are real good 'uns, and are very fit, and still look in splendid condition. There are several waggons outspanned near us. Feathered game was scarce to-day. Inspanned at 4 p.m. Soon after starting Giles received a note from Dr. Saur and Mr. Williams, mining engineer, who are going up to Mashonaland for the Zambesi Exploring Company. They asked for the "loan" of some bread and a few necessaries of life. It appears that they have been coming up the road quickly in a Cape cart, and expected to catch up their waggons about here. Unfortunately their waggons had by accident taken a wrong road, and they were stranded without any "skoff." Of course Giles soon found them the necessaries required.

Wednesday, June 24th.—Inspanned at 1 a.m., and at daybreak arrived at Silika, twenty miles from our last camp. The road was very rough. There used to be a store here, but it has been moved. There is a small detachment of the Bechuanaland police. We have left the Crocodile River well to the right. This is a very prettily-situated place. There is a large kopje at our back and several others around. There is a small stream of running water about half a mile distant. The outspan place is very dirty. There are lions about here, and a Kaffir shot one the other day and sold the skin to a white man for 15s. There are also koodoo and giraffes. Dr. Saur and Mr. Williams arrived in their Cape cart, and were made honorary members of our mess. They had

The main column encamped on the bank of the Lotsani.

seen a herd of wildebeest just a few miles away, and Mr. Williams, while looking for feathered game, came across a hyæna, which he immediately let drive at and killed. We rest here to-day, as there is a twenty-four mile trek to the next water.

All along the road for the last few days we have come across dead bullocks, the result of lung sickness.

Thursday, June 25th.—Inspanned at 2 p.m., and at 4 we outspanned for an hour. Outspanned again about 8, having trekked about 12 miles, Dr. Saur and Mr. Williams following in their Cape cart. John (our cook) has been seedy with a bilious attack, and Mackay had a headache—the result of a bathe, which he, Edgell, and myself took in a nice clear pool which we found this morning. The water was rather cold. The dust on the road seems to get worse every day. It gets into one's mouth, eyes, nose, and ears; fills one's kartel, and makes everything filthy. It is always red sandstone, I suppose.

Friday, June 26th.—Arrived at Lotsani River at 9 a.m., which we crossed and camped on the further side. The road through the river was down and up steep banks, but we came through it well. We have done twenty-five miles from Silika in three "skoffs,"[1] which is excellent trekking. The river is very low, but the water is clear. It is, however, brackish and unpalatable, and is apt to produce diarrhœa. The outspan place is dirty, but none other is possible. The nights have been much warmer the last few days, and it is no longer a question of sheepskin kaross and numberless blankets. To-day we discovered a Hindoo barber, who is making his way up country on some Kaffir

[1] Skoff; journey from outspan to outspan, or from meal to meal.

waggons, so we all indulged in the luxury of a cut and shave. Inspanned at 4 p.m., and trekked six miles, that is, about two miles beyond Elebi, crossing a small drift on the way. Elebi is a small police-station, there being now two men there.

The camp of the main column at Suchi River.

The fort that was here has been abandoned. The place is of some importance as one where police can be concentrated for patrolling the Crocodile River, which is about twelve miles distant, in case of trouble with the Boers.

Saturday, June 27th.—Arrived at Suchi River at daybreak, and encamped on the other side. Our trek here was about sixteen miles. The River Suchi is similar to the Lotsani, being now merely a thread of water in the river-bed. The water is brackish and unpalatable, and leaves crusts of salt on the banks where it has evaporated. The country is flat all round. About ten o'clock this morning it commenced to rain, and rained in showers for about an hour. Thunder was heard in the direction of the Transvaal. Rain is a very unusual occurrence in these parts at this time of the year.

Tuesday, June 30th.—Arrived at sunrise at Macloutsie, and camped on the other side of the River Matlaputta. Macloutsie consists of a police camp, telegraph station, two good stores, and a fort. It is now the headquarters of the Bechuanaland Police. There are at present about 100 men stationed here. About two months ago the telegraph wire was prolonged here from Pelapswe, and since then on to Tuli Fort. The horse sickness is very bad here, and I was informed that about 80 per cent. of the horses died last year. The river is small, but the water is very good. Giles, Edgell, Mackay, and I rode in and called at the store, where we bought a set of cricket materials. It seemed odd to find such things for sale in the midst of an African wilderness. In the evening we dined at the officers' mess. After dinner the band, which consisted of a violin, a flute, and a guitar, played, and we passed a very pleasant evening. Such a charming, cheery lot of fellows, and most hospitable too. The officers all live in thatched huts, and the mess hut is the same sort, but on a larger scale, of course.

Wednesday, July 1st.—Trekked to the further side of the Macloutsie River, about five miles. At present the river is a small stream of good, clear, running water, and about eighty yards in breadth at the crossing. The descent and ascent are fairly steep, and it is a stiffish pull for waggons. Some Kaffir waggons following us that are carrying ammunition, etc., to Mashonaland had to double span each waggon, and then they had a lot of trouble because their trek chains broke over and over again. We trekked about four miles after sundown. Road was very hilly and crossed by many dry spruits.

Thursday, July 2nd.—Arrived sunrise Lipokwe River; good road from our last camp, which is about eight miles away. River now about five yards in breadth; clear, good running water. There are many pheasants and guinea-fowl here, and our larder is now well supplied with game.

Friday, July 3.—Arrived Semalali River, about eighteen miles trek. I don't think I am likely to forget this place. It is the easiest thing in the world to lose one's way in this country, and to-day is not the first time it has happened to me. You take careful landmarks of kopjes, the direction of the wind, the position of the sun, etc.; you provide yourself with a pair of field glasses and a compass, and then imagine that it is impossible to mistake the direction from which you came. And yet when you arrive at some point to which you have taken a bee-line, say a couple of miles away, you look back, and, somehow or other, the whole scene seems changed. Your landmarks appear in a different position, the wind is now in another quarter, and your camp, from which you could see distinctly the spot on which you now stand, is invisible. You search the landscape carefully with your field glasses, and all looks different. That hill over there should be more to the right; that other smaller one should be more to the left and nearer. You are loth to believe at first that you do not quite know where you are, but as you walk on, thinking you are going in the right direction, your landmarks become more and more changed. All around you is a boundless stretch of undulating plains covered with bush and scrub, sometimes so thick that you see nothing beyond fifty yards. Occasionally you come across a kopje, when you have no idea you are anywhere near one. Not a sound is to

The waggon conductor sports a new pair of "store" trousers.

be heard, except perhaps the occasional twittering of a bird or the rustle of the leaves and long grass. At length you feel obliged to own that you don't know where you are. It is a time of desolation, and you cannot but feel how utterly helpless you will be should you be unable to find your camp before sundown. It was this feeling that I experienced to-day. I went out soon after 8 a.m., having taken a little coffee and biscuit, only intending to potter about after pheasants and guinea-fowl for a couple of hours or so. I crossed the river and walked towards a small kopje. In about an hour I thought it time to return for breakfast; but, somehow or other, missed my way in some long grass. I thought it didn't matter very much, as I knew the general direction of the road, so steered north-west so as to cut it at right angles. But I walked on and on through the wilderness, and no road appeared. After more than a couple of hours' hard walking in the hot sun with three dead guinea-fowl dragging on my waist-belt, and a heavy gun, which felt heavier every moment, on my shoulder, I came to the conclusion this wasn't good enough, and determined to break a rule which I have often had instilled into me in this country, namely, that when once you strike out in a certain direction you shouldn't go back. However, I am glad I did go back, for I know now that the road at this particular point goes south-east, or nearly so, whereas all the way up to now it has, of course, been north-east. The long grass was very trying, and I never was as thirsty in my life. A Yankee can boast about a ten-dollar thirst, but I'd have given mine away for nothing, and I never want another one like it. Suddenly I heard a rustle, and, looking up quickly, saw a herd of splendid hartebeest, which animal I had not seen before, and which I recognized by their horns. There were nine of them, and they came along at a sort of canter trying to head me to windward. They stopped all of a sudden at a little over 100 yards, offering a splendid shot broadside. Alas! I had only a shot-gun with No. 5 shot, but I couldn't help going down on one knee, and taking aim at one just behind the shoulder, and thinking how I must get him with rifle. Off they went again, and I was alone once more in the wilderness. About five minutes afterwards I came upon a pool of beautifully clear water—what was left of a small dried-up stream—and fairly wallowed in it. While here I thought I heard three shots fired at distinct intervals, so, knowing that was the signal agreed upon in case any one lost his way, struck out in that direction. Then from the top of a kopje I made out the river, as I thought, by a line of green trees. This proved to be correct, and I then soon came up with our own oxen grazing. The boys directed me to go along the river bank back to camp, saying I couldn't miss the way. By accident I was told the wrong side of the river, so after walking about three more miles, I managed to lose myself again, as the river all seemed to go to nothing, and I couldn't tell which was river and which was veldt. Therefore

SHOWING A FLARE-UP FOR THE LOST ONE.

Page 140.

I walked back, thinking to find the oxen once more; but the sun was getting low, and I found them gone, and, worse still, couldn't trace the spoor. Then I heard shots fired, and going in the direction of the sound, came up with a Kaffir with some oxen, who showed me our waggons about five hundred yards distant. I got in after sunset, and found the waggons just gone—all except the mule waggon, which was waiting for me. Thank goodness! I'm here at last. I've walked hard from eight this morning till sundown without a morsel of food. I didn't quite relish the idea of sleeping out in the cold veldt with nothing on my shoulders but a flannel shirt, and no fire — for, *mirabile dictu*, I had forgotten cigarettes and matches. In ten minutes we caught up the other waggons and trekked eight miles. I forgot to say that the three shots I thought I heard were the three signal shots, sure enough, from our waggons, and they probably saved me several miles walking. It only shows how useful it is to have a signal agreed upon. One thing I am certain of—the man who says he can't lose himself in this country (and I heard one once) is a fool. Nothing is easier.

CHAPTER X.

TREKKING AND HUNTING.

We entertain Sir Frederick Carrington—Farewell to Fort Tuli—The business of inspanning—Our camp at night—Sport with Dr. Rayner and Lee—Laying the telegraph wire—The Umzingwani River Camp—Koodoos, quaggas, and honey birds—Lee's boy nicknamed "The Baboon"—The elephant fruit-tree—Lee a charming companion on the Veldt—The Umsajbetsi River—Habits of our oxen and mules—Shooting game in South Africa—A native market—An unsuccessful antelope hunt—The mahogany tree—Further hunting experiences—Camp on the Bubjane River—Our conductor Myberg.

On the evening of July 16th, our party entertained Sir Frederick Carrington and some of the officers of the B.B.P. at a farewell *al fresco* banquet, and passed a cheerful evening round the camp fire with its usual accompaniment of song and tale. I had accompanied Sir Frederick in the afternoon on a shooting excursion after reit-buck. These buck were expected to be found in a long and wide glade near the Limpopo, where the rushes were high and the grass was thick. Some dozen mounted troopers acted as beaters, and we saw seven buck, of which two were killed. They are about the same height as a fallow deer, with red bodies and white bellies; their horns are short and pointed. We also secured two brace and a half of pheasants. On the 17th, the waggons with our baggage from Pretoria having at length arrived, we left Tuli.

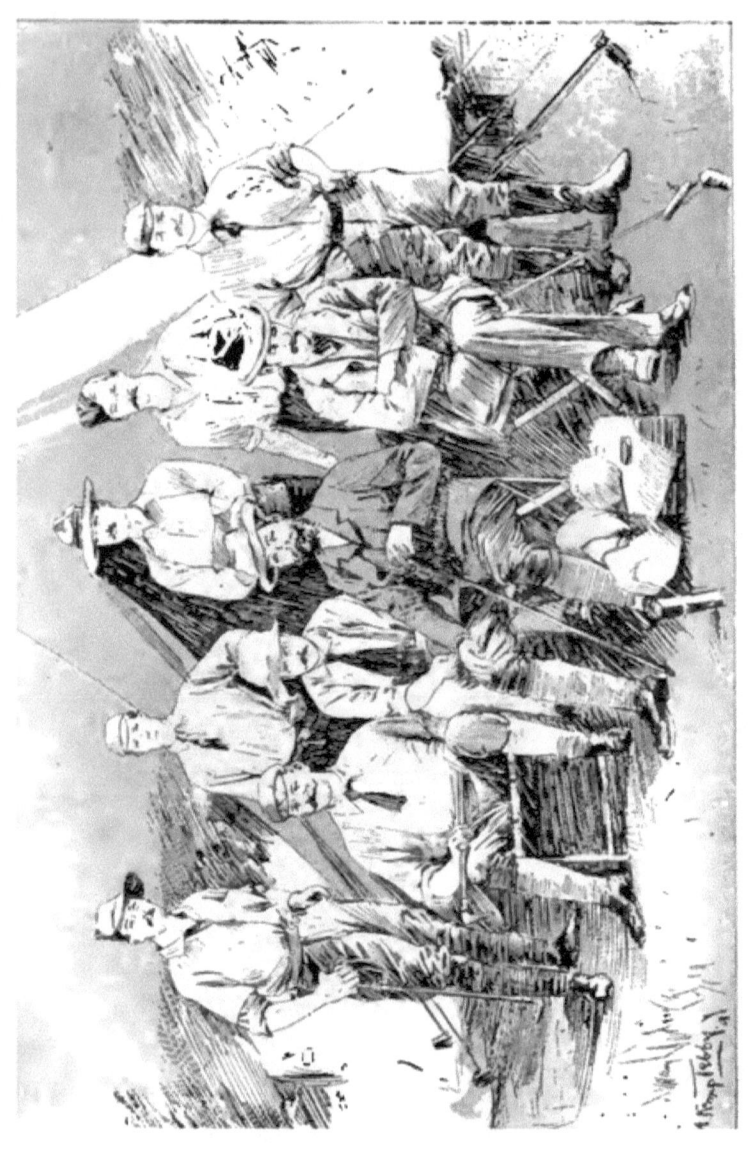

MR. MACKAY.   MAJOR G. GILES, R.A.   LORD RANDOLPH CHURCHILL.   MR. H. C. PERKINS.   SURGEON HAYNE.
CAPT. G. WILLIAMS.   (Mining Expert).   CAPT. COVENTRY.

THE MEMBERS OF THE EXPEDITION.

Mr. Alfred Beit left early in the morning of the same day, his waggons having preceded him some twelve hours. The business of inspanning, when a novelty, is very interesting. The camp presents a scene of great apparent confusion, but in reality all is in perfect order. The various cases, portmanteaux, and bags, having been packed, the tents struck and rolled up, and the bedding folded, and everything being assigned to its proper waggon, the loading of the waggons begins, a work requiring great care and method. All this work is done under the orders of Mr. Edgell and Mr. Mackay, whose task is by no means a light one. The marshalling of over a hundred oxen, of the horses, mules, and donkeys, proceeds with precision and regularity, the "boys" having been perfectly drilled and trained on "the trek" through Bechuanaland. All being ready, the "vorelopers" at the head of their teams, the drivers causing their whips to crack with loud reports, off starts one of the waggons, five minutes later another, and so on; last comes my "spider" with its team of eight mules. The whole made a fine procession of great length. At the outset a work of difficulty lay before us, the crossing of the drift of the Tuli River. Here the sand for more than one hundred yards is deep and heavy, and double spans become necessary for each waggon. The leading waggon, having descended into the river-bed, is halted, the span of oxen is taken out of the second waggon and attached to the first, which, drawn by thirty-six oxen, moves with apparent ease through the drift. This process, repeated with each waggon, occupied some two hours, and it was four o'clock before all the waggons were safely over on the other side. The seven waggons, the Scotch cart, and spider, all crossed over without the slightest stickfast, accomplishing what I was informed was a record achievement. Sir Frederick Carrington and some of the officers of the B.B.P. watched our proceedings, and no doubt if there had been any hitch, or if any of the waggons had stuck, much chaff would have been indulged in at the expense of the expedition; but the latter, stimulated by the knowledge that critical eyes were looking on, were resolute to prevent the smallest mishap. After trekking six miles we outspanned and set up our camp for the night. The appearance of the camp was striking. The moon shining brightly, the long avenue of waggons on each side of the road with the oxen lying down, attached to their yokes, offered a most singular and memorable sight. I had had a couple of hours' shooting in the afternoon with Sir Frederick Carrington, and brought into camp a small roi-buck, a hare, and two pheasants. On the 18th every one astir by half-past five, the waggons were started off at daybreak. They trekked seven miles, outspanned at nine, the sun being already warm. Dr. Rayner, Lee, and I rode into the bush to look for buck. I found one lying dead in a small pool of water, which had been shot the day before by some unfortunate sportsman. We carried it off in triumph to the camp, which we reached about eleven o'clock. Washing, breakfast, and loitering about occupied

FIRST NIGHT OUT FROM FORT TULI.

the time till four o'clock. It is but true to say that I was the only loiterer, every one else having some kind of work to do. Our day camp was pitched on the Ipagi River, where were also encamped a large body of men employed by the Chartered Company in laying the telegraph wire up country to Fort Victoria, which work is being accomplished at the rate of about three miles a day. This expedition is mainly composed of 250 of Khama's men, all armed with old muskets, which they carry slung over their shoulders, generally loaded and at full cock, together with their picks, spades, and axes. It is doubtful whether Lobengula will quite relish the incursion into his country of armed men from the tribe of his hereditary foe. Khama once, some years ago, nearly killed Lobengula, wounding him badly in the neck. The work, however, of laying the telegraph has to be done, Khama's men were the only labour which could be obtained, and Khama's men would not come into Lobengula's country unless they were fully armed.

On the 19th we reached the Umzingwani River, about twenty miles from Fort Tuli. There was a good amount of water here. Major Giles, I and Lee, after breakfast, rode out into the bush to look for game. Two water-buck were seen, which Lee shot at without effect. I got a shot at a steimbuck, but missed him. We saw much spoor of koodoo and other antelope. A heavy shower came on, a very unusual thing in this part of Africa at this time of year, and we got wet through. It was now found that our hours of trekking, which had been

L

adopted mainly on my account, were unsuitable for the oxen. To make good treks it is necessary for the oxen to labour either during the night or late in the afternoon when the sun is low. We decided to return to the hours of trekking which had been adopted by the expedition during their passage

Our camp on the Umzingwani River.

through Bechuanaland, from 1 a.m. till daybreak, and from 4.30 p.m. until 8 p.m. When the waggons started in the early morning my "spider" remained behind till 6 a.m., and caught the waggons up by breakfast time. On the 20th I rode with Lee into the bush. We came across two koodoo bulls, one of which Lee shot. The koodoo is a magnificent antelope. It stands as high as a mule, is of a soft grey colour, its face is beautifully marked with white, and it carries fine twisting horns from two to three feet long. Further on we put up two wild pig; Lee got one and I got the other. We saw much fresh spoor of quagga. This morning I saw for the first time the honey-bird. We followed it for about half a mile. When Lee whistled, it gave back an answering note, flying from tree to tree, leading us on. When it reached the tree occupied by the wild bees it answered no more to Lee's whistle, indicating that the honey was found, and flying off to a neighbouring tree to watch our proceedings. As we were unprovided with an axe, the poor bird was destined to be disappointed in us. Lee and his boy both climbed the tree, found the holes into the hive, and got well stung. Lee's boy is a most remarkable creature. He is a long-legged, lanky bushman, answering to the name of "Baavean," pronounced "Bobean," the Dutch for baboon. The "Baboon's" skill in spooring game is almost incredible, he possesses an instinctive knowledge of the habits and as to the whereabouts of animals. Lee and the "Baboon" will spoor game through the bush for miles. A tree pointed out to me this morning, the "elephant fruit tree": bears a small fruit about the size of an apricot, from which when ripe exudes an amber-coloured syrup, which tastes when eaten something like a preserved candied greengage. Elephants are said to be very fond of this tree, from whence comes its name. Lee I find an excellent companion on the veldt, for, besides his great shooting skill and experience, he possesses a large amount of bush lore in respect of animals, of trees, and of plants, which he imparts freely and agreeably. These morning rides through the bush have an indescribable charm. The scenery, the fresh air, the bright sunshine, and the knowledge that you may at any moment come upon anything in the shape of game, from a lion or a giraffe down to a pig or a baboon, lends to these excursions a most exhilarating interest. We rejoined the camp about midday on the Umsajbetsi River. From this river to the Umzingwani is a long stretch of seventeen miles without water for the oxen. The Umsajbetsi at this time of the year is only a bed of dry sand, but water somewhat brackish is easily obtainable by digging a foot deep in the sand. Captain Williams went out shooting in the afternoon, and wounded badly two koodoo cows, but unfortunately both got away. At five o'clock we inspanned. I find it very amusing to study the habits of the oxen. In spite of their long horns and somewhat wild, formidable appearance, they are, in reality, to those who have to drive and manage them, the most docile, patient animals. A stranger, however, would do well to be careful not to go too close either to them or to the mules. These oxen come in in the evening from the veldt in one great troop, driven along by a couple of boys. They range themselves in spans, as schoolboys at a school range themselves in classes, each span apparently knowing its own waggon, each ox its own place in the span. The mules are not so interesting or attractive, and it is possible that a mule is one of the few animals on which kind treatment is absolutely thrown away. Our mule waggon, which loads over 2000 lbs. of transport, has a fine team of twelve mules. They are a most vicious set, and would readily bite or kick at any one except Myberg, the conductor, or Gideon, his "boy." Myberg tells me that they would even go at him if he happens to wear a different hat or coat from that which they are accustomed to. These mules have their idiosyncrasies. One of them is that they like to be accompanied by a mare. This mare is tied up alongside the span, but does no work herself. She goes out grazing with them on the veldt, and I am told that when mules have a mare along with them they never stray. Another curious habit of theirs, which it often amuses me to watch, is that of gnawing each other. This gnawing appears to be a regular matter of bargain between them. Two mules approach each other, one wants his shoulder gnawed, the other his quarter. Their conformation makes it necessary for the proceeding that each should gnaw the same place on each other at the same time. The mule with the itching shoulder suggests to the mule with the itching quarter, "If you will gnaw my itching shoulder for a few minutes I will gnaw your shoulder which does not itch, but will then gnaw your quarter which does itch, and allow you to gnaw mine which does not."

Just before inspanning this evening the dead

koodoo was brought in on two donkeys, and was the object of much admiration, and from no one more than the cook. On the 21st we reached the Umshlane River, a ten miles trek. This river is also dry, but water can be got by digging. Having found that game in the vicinity of the road was scarce, probably frightened away by the constant passage of waggons and by the telegraph expedition, I arranged to go with Lee for three or four days away from the road into the veldt. I took with me the "spider" and the Scotch cart, a small two-wheeled waggon, and provisions for six days. The mules were taken from the mule waggon and attached to the Scotch cart, and the oxen from the Scotch cart were put to the mule waggon. I started off in the afternoon and reached the Bubye River at sunset. The next morning at daybreak we rode off into the veldt just as our waggons, which had been trekking through the night, passed us. Soon we came across the spoor of koodoo and quagga mixed, which Lee and the "Baboon" followed for upwards of half-an-hour. A low whistle from the "Baboon" denotes that he perceives the antelope. I jump off my horse and see through the trees very indistinctly three koodoo about 150 yards off, at which I fire without success. They gallop off, and we follow on their spoor, and come suddenly upon some roan antelope at about the same range as were the koodoo. Again I jump off my horse and fire, and again without result. This South African shooting is a widely different business from Scotch deer-stalking. In Scot-

EXPERIENCES OF SOUTH AFRICAN SHOOTING. 151

land one is taken by some steady old stalker within a hundred yards or so of the stag, which is generally standing in an open space, and offers you a fair "pot-shot," the rifle being rested either

Typical natives from the Umshlane River districts.

on your knees or on a stone. Here you have to jump off a horse, look through a quantity of trees and bushes, fire from the shoulder, and fire quickly, as the game nearly always sees you as soon as you see it, and bounds off. Moreover, it is incredible how difficult it is for an untrained eye to discern these wild animals through the bushes. On more than one occasion, though I possess a tolerably good pair of eyes, Lee has tried in vain for some seconds to show me antelope through the bushes, which I have been totally unable to make out. I expect that to be successful after game in the South African veldt requires long training and experience. Lee galloped away after the roan antelope through the bush, helter-skelter. I remounted and followed him as best I could, but lost sight of him. I heard Lee fire three shots, and, on coming up with him, found that he had got one antelope on the ground, about 80 yards off, and another badly wounded about 100 yards away, moving off. We followed up the wounded one, and perceived that it was accompanied by another buck, who was apparently unwilling to leave it. I get a good shot at this one, and kill it, my bullet passing through both shoulders. The wounded one is finished off with another shot, and there are now lying on the ground within the space of 200 yards three roan antelope, a big cow with splendid curving horns, and two young bulls whose horns were shorter and almost straight. Truly this was a fine sight, and one which some English sportsmen would gladly travel 8000 miles

MARKETING WITH THE MAKALAKA.

to see. The roan antelope is rather smaller than the koodoo, about as big as a fine Scotch stag, and quite as graceful in appearance. The main body of our waggons was only encamped about three miles off. The "Baboon" is at once sent off to fetch donkeys to carry the meat, while Lee and I remain to grallock the bucks and cover them over with grass and branches to hide them from the vultures. We then rode on to Major Giles' camp, where I luckily found my friends at breakfast. Here I was informed that the eight oxen were not strong enough to drag the mule waggon, so it was decided that the mule waggon was to remain behind with me, and the light Scotch cart was to be given back to the span of oxen. This arrangement was a pleasant one for me, for by it I obtained the companionship of Captain Williams, who occupied the mule waggon. I remained with my friends for some hours, and witnessed for the first time a regular native market. A small group of Makalaka had a kraal on the Umjinge River, close to our camp, and brought pumpkins, milk, mealies, and beans, for which they took in exchange pieces of coarse blue calico ("limbo"). Trade proceeded merrily, with much laughter and joking. Mr. Mackay and Mr. Coventry conducted the barter, but I am afraid that their weights and measures would not always have sustained the examination of an English police inspector. I found that one yard of "limbo" would purchase about a shilling's-worth of stuff. Altogether some twelve yards were expended. These natives were by no means an unattractive lot, some of the women having a bright and youthful appearance which rendered them almost good-looking. They were very partially clothed, but much adorned with feather coiffure, and brass ornaments on arms and legs. I rode afterwards with Lee to make a circuit through the veldt back to my camp. We had not proceeded far into the veldt before Lee pointed out to me, about eighty yards to my left, a sable antelope. This magnificent creature, with long horns arching right over on to its back, was standing in some high grass looking at us curiously. On horseback I saw him perfectly, but when I jumped off to fire I could only see the top of his head and his horns, owing to the long grass. I took a careful aim through the grass at where I thought his shoulder ought to be, but, alas! without effect. He bounded off, Lee in hot pursuit. Lee got a shot at him some distance further on, but missed. We followed him, and came upon him a third time, but got no shot, as he was too quick for us, and made off for good. We were rather unhappy over this reverse, for the sable antelope is the antelope of all others which the South African hunter covets. In the course of the afternoon we saw a great deal of spoor of various kinds of game, but got no further shot. On my return to camp I found that Captain Williams had arrived with the mule cart, and great plans were made over dinner as to future sport.

The following morning we were both off at dawn, Captain Williams going in one direction accompanied by the "Baboon," Lee and I proceeding towards Mount Towlu, which rose from the plain in solitary grandeur to an altitude of about 5000 ft. above the level of the sea, six miles distant from our camp. I passed a profitless day, seeing nothing but a couple of koodoo cows, at which Lee got a snap-shot. I saw, however, some objects of interest. We came across a fine big mahogany tree covered with seed pods. These seed pods resemble a very large locust bean, and the covering is like old black shoe-leather. On opening them you find, arranged in beautiful order, about eight or nine seeds in shape like acorns, the cup being the brightest scarlet, the berry ebony black. Crossing a sandy patch, Lee pointed out to me the spoor of two lions, which he said was about two days old. I now began slightly to realize that one might come across a lion some of these shooting days, as to which I had been for some reason or other rather incredulous. Thirty-six hours from this time I was destined to have all doubts as to the existence of lions dissipated in a startling and not altogether agreeable manner. I got back to camp very tired about four in the afternoon, and found that Captain Williams had been equally unsuccessful, having only had a long and difficult shot at a hartebeest, and a bootless chase after a sable antelope. Our ill-luck did not prevent us from making an excellent dinner off stewed roan antelope of our own cooking. During our absence our camp had been moved in an easterly direction, some three and a half miles away from the road into the veldt. Considering that it was still too near the road to see much game, we determined to devote the next day to moving the camp a good distance further into the veldt down the Bubjane River. This we effected. It was a trek of much interest, as we had to make our own road through the bush. Captain Williams, I and Lee kept riding on ahead to find out the best way through the trees, the places where the many spruits could most easily be crossed, and how to avoid the rocky ground. We had to cross the river Umjinge just where it joins the Bubjane. This was rather an anxious business, for the banks were high and the sandy bed was heavy. The mules were taken out of the "spider," and attached to the team of the mule-waggon. Drawn by eighteen mules, this waggon literally bounded down the bank, more than once within an ace of capsizing, and for a moment stuck fast in the sand. Our mules, however, proved equal to the emergency, and, stimulated by the most tremendous cracking of "vorschbachts" and some lashings, successfully dragged their load up the opposite bank, where they were soon followed by the "spider." At ten o'clock we outspanned and rested till one in the afternoon, when we inspanned again and trekked till sunset. We accomplished altogether about ten miles. The axe had often to be freely resorted to to clear away the bush ahead. We pitched our camp in a lovely spot on the high, precipitous bank of the Bubjane River, which cannot be said to flow, but which lies below us in a series of pools of clean and clear water, dotted here and there among huge boulders of rock and wide spaces of sand. All around us is the thick bush veldt. We have the place all to ourselves. In the distance Mount Towlu, behind which the sun sets with a scarlet glow. The moon, almost at the full, illuminates the surrounding scenery with astonishing brilliancy. We were a cheerful party that night at dinner, Captain Williams and I, Myberg the conductor, and Hans Lee the hunter. Myberg is a splendid specimen of a young colonial. Standing six foot five in his stockings, strong as a horse and wiry as an antelope, he possesses a most good-natured disposition, is always ready for anything, and makes the best of everything. Hans Lee is a short but well-made man, with regular features, a black beard and moustache, a soft droning kind of voice which lends to his conversation and his narratives a peculiar charm. His English is rather broken. We retired to rest early, somewhat fatigued with the labours of the day; the sounds of the night were the crunching of the mealies by the mules attached to their canvas manger, their whines when biting each other, the occasional howl of a jackal or a hyæna, and at 1 a.m. the "Baboon" woke up Captain Williams to make him hearken to the roaring of a lion some three or four miles away on the other side of the river. The sound, it appears, was quite faint, and I was somewhat incredulous when told about it next morning. By ten o'clock I was perfectly convinced of my error.

CHAPTER XI.

LIONS.

Lion Camp—The Tales of a Huntsman—The snake-tree—In the track of the koodoos—We come across a posse of Lions—Antelopes and quaggas—Return to camp for the dogs—Result of one day's sport—We spend another day hunting—Provisions running short.

ON the evening of our arrival at "Lion Camp," while Captain Williams and I were preparing the dinner, Lee had gone out on foot with his rifle, and had shot at and wounded a sable antelope only a short distance from the camp. This buck we started to hunt up on the morning of July 25th. Lee and the "Baboon" soon found its spoor, but were unable to follow it far, and we shortly afterwards gave up the pursuit. We then separated, Lee and I hunting towards the east, Captain Williams and the "Baboon" going south. On this day we were poorly provided with horses for hunting purposes. I had left my shooting pony "Charlie," a perfectly-trained and steady animal, from off whose back I can fire, at the camp, as I had ridden him the previous day, and I was mounted on a strong, somewhat under-bred bay horse, a good roadster, but unsuited to shooting or to the chase. Lee was riding a weedy little chestnut pony, which had been purchased at Kimberley for the sum of 8*l*. Captain Williams was mounted on a bay pony named the "Tortoise," which name sufficiently describes him, and the "Baboon" bestrode a large raw-boned, cock-throppled nag called "Nelson." But it had never occurred to Captain Williams, nor to me, that anything very wonderful in the way of steeds was necessary. We looked upon them merely as conveyances for getting over the ground quicker than we could on foot. Since this morning I have come to the conclusion that shooting in South Africa, unless it is to be accompanied by great risk, requires that the sportsman should be mounted on a perfectly-trained, well-bred, fast horse, just as tiger-shooting in India requires a perfectly steady and courageous elephant. For some time, nearly two hours, Lee and I wandered on, peering about through the bushes, examining spoor, of which we saw much, and sometimes conversing in a low tone. At times Lee told me of one or two curious things. He told me that in the Zoutspansberg round where he himself resided, there grew a tree called the snake tree, the leaves of which, when boiled, make a decoction which is an infallible specific against snake bite. He had used it himself, he said, on animals with complete success, and he was perfectly confident that with this remedy he could cure any human being who had been bitten by a poisonous snake. He further told me that persons whom he could trust had informed him that the drink made from these leaves cured animals from the mortal illness caused by the bite of the Tsetze fly. This is a matter worth examining into, as Lee is a person of perfect credibility. He also informed me that there was a small tree in Matabeleland, where he has lived for many years, which bears a fruit like to and sweet as the pine-apple, the roots of which are a perfect antidote to the effects of strychnine poisoning, and are always used by the natives to cure any of their dogs which have picked up poisoned food which has been laid about to kill jackals and hyænas. Also he recounted to me other stories of his experiences in the chase. At last we perceived, some distance off, two koodoo cows. We approached them within tolerable range, and both dismounted to shoot. The cows were making off. I dismounted so awkwardly that I fell heavily on my back, cutting my hand and losing my hat. Getting on and off a horse with a heavy rifle in your hand requires practice like anything else. Lee got a shot and struck one of the koodoos, and galloped after it. I picked myself up as well as I could and followed. On arriving at the place where the koodoos had been the spoor was examined, and much blood was found on the grass. We followed the track of the wounded animal for some space, but had to give it up. It was now past ten o'clock. The sun was high and hot; we had seen little, and I began to think that I was going to have another day barren of sport. Lee climbed up a kopje, beneath which we were riding, to examine the surrounding bush, and after about a quarter of an hour's absence rejoined me, and said he could see nothing, but thought we had better turn to our left towards the north, as in the direction which we were taking there was nothing but thick bush, whereas towards the north the veldt was much more open. In a few minutes I almost wished that we had stuck to our original direction.

We were riding along through a small open glade covered with high grass, Lee a few yards ahead of me, when I suddenly saw him turn round, cry out something to me, and point with his finger ahead. I looked, and saw lolloping along through and over the grass, about forty yards off, a yellow animal about as big as a small bullock. It flashed across me that it was a lion, the last thing in the world that I was thinking of. I was going to dismount and take aim, but Lee called out in succession five or six times, "Look, Look!" at the same time pointing with his finger in different directions in front. I saw to my astonishment, and rather to my dismay, that the glade appeared to be alive with lions. There they were trooping and trotting along ahead of us like a lot of enormous dogs, great yellow objects, offering such a sight as I had never dreamed of. Lee turned to me and said, "What will you do?" I said, "I suppose we must go after them," thinking all the time that I was making a very foolish answer. This I am the more convinced of now, for Lee told me afterwards that many old hunters in South Africa will turn away from such a troop of lions as we had before us. We moved on after them a short distance to where the bush was more open, the lions trotting along ahead of us in the most composed and leisurely fashion, very different from the galloping off of a surprised and startled antelope. Lee now dismounted and fired at a lion about fifty yards off. I saw the brute fall forward on his head, twist round and round, and stagger into a patch of high grass slightly to the left of where I was riding. I did not venture to dismount with such a lot of these brutes all around ahead of me, not feeling at all sure that I should be able to remount quickly enough and gallop away after shooting. My horse, untrained to the gun, would not allow me to fire from his back, and would probably have thrown me off had I done so. I stuck close to Lee, determined to leave the shooting to him unless things became critical, as his aim was true. I counted seven lions; Lee says there were more. I saw, and cried out to Lee, pointing him to a great big fellow with a heavy black mane trotting along slightly ahead of the rest. He was just crossing a small spruit about 100 yards ahead, and as he climbed the opposite bank offered his hind-quarters as a fair target. Lee fired at him, at which he quickened his pace and disappeared in front. We approached the spruit, and, almost literally under my nose, I saw three lions tumble up out of it, climb the opposite side, and disappear. Now I own I longed for my shooting pony Charlie, for they offered me splendid shots, quite close, such as I could hardly have missed. I raised my rifle to take aim at the last, but, perhaps fortunately for me, he disappeared, before I could fire, in the high grass on the other side. I saw Lee fire from his horse at one as it was climbing the bank, which he wounded badly, and which retreated into a patch of thick grass the other side of the spruit, uttering sounds something between a growl, a grunt, and a sob. The lions had now got some 100 yards or so ahead of us, and had disappeared into thick high grass. We knew that there was a wounded one behind us on our left, and another wounded one in front of us also on our left. Lee now got rather excited. I have no doubt that if he had been by himself, mounted on a good horse, he would, to use his own expression, have "played the devil with them." He told me that lions would not stand being chased very far, but would lie down, conceal themselves, and wait for us, and that if we approached the wounded lions they would in all probability charge, when we should have to gallop away at the top of our speed. The idea of galloping at full speed on a second-rate horse through thick bush trees and grass, chased by a lion, was singularly unpleasant to me. After a few minutes' consideration, and after making me promise to remain where I was and gallop away as fast as I could as soon as he had fired, Lee determined to go and look at the second wounded lion, who was lying away from us some sixty or seventy yards. I saw him go up to within about ten yards of the grass where the wounded lion lay. Fortunately she was badly wounded by the first shot, or she would probably have come at him. He caught a glimpse of her and fired. There was a tremendous commotion in the grass for a second or two. Lee's horse gave such a bound that I thought he would have been thrown, turned round, and galloped away. I followed as fast as I could. We galloped about 200 yards to the right of the direction the other lions had taken, and then pulled up and held another hurried and anxious consultation. Lee wished to go back straight to camp, about three miles off, and get our three dogs, two pointers and a greyhound, which he said would soon show us where the lions might be lying, as in the thick grass we could see nothing. I was anxious to make certain of the lion last fired at, feeling sure that it was mortally wounded, and could do no harm. And now occurred a strange thing. Six koodoo cows came suddenly galloping along fifty or sixty yards away from us one after another. I called out to Lee not to shoot, as they were only cows, without horns, and I did not want matters still further complicated. However, he was not to be denied, jumped off, and shot at the last koodoo, bringing her heavily to the ground. She got up again and made off. In two minutes the koodoos, which had been galloping in the direction where the lions had last been seen, came galloping back past us upon their tracks, showing pretty clearly that they had gone right upon and had winded the lions, who were lying about near their wounded fellows. Lee now insisted that we should go straight back to the camp and get the dogs, as the situation, he declared, was one of danger. I confess that when I was a quarter of a mile from the spot I felt rather relieved, for I had thought all along that eight or nine lions was *trop de luxe*. As we were going along Lee exclaimed, "By Jove, there's a lot of buck!" Away he galloped, and I after him. We came pretty close up to a lot of about a dozen roan antelope and three quaggas. I dismounted and fired at one, with what effect I do not know, for Lee galloped on, calling to me to get on my horse again. I got up and galloped after him; he being a considerable distance ahead, I had some difficulty in keeping him in view. I heard him fire three or four shots, and, when I got up to him, found that he had one antelope apparently dying on the ground, and two more staggering away badly wounded. Just now three antelope cantered by to my left, and I got a capital shot in the open, about seventy yards. I hit one very hard with my first barrel, but did not stop it, caught my horse with a little difficulty, and galloped on after Lee, who was again a good way ahead of me. The place seemed alive with game. I came up to Lee, who had again dropped another antelope. I saw a quagga about eighty or ninety yards off, fired at him and dropped him just a little beyond Lee's wounded antelope. We walked on towards the antelope and quagga,

Lee giving them each a finishing bullet. We then remounted and galloped on, Lee shot at another antelope and wounded it severely, and I fired at and broke the hind leg of a quagga. The game all made off as best they could, and I could do no more, being perfectly exhausted. I had dismounted and fired seven times, and galloped very hard for nearly two miles. We "off-saddled" and had a little breakfast, of which whisky-and-water was the principal element. The sun was very hot. Lee was certain that we should pick up five or six antelope and two quaggas, but these hopes were ultimately disappointed. After a hurried repast we got on our horses and rode to the camp, within a mile of which we fortunately found ourselves, arriving there about midday.

Captain Williams and "the Baboon" had not returned as we had hoped; so, after waiting for them in vain for an hour and a half, we started back again to hunt up the wounded lions, taking with us Myberg, armed with a smooth-bore and ball cartridge, and the three dogs. My servant Walden being very anxious to accompany us, after some hesitation I permitted him to come on condition that when we approached the place where the lions were he should ascend a tree. He was armed with a Martini-Henry rifle, and led the dogs. I had no horse for him to ride. I thought we were going to have a nasty business, for Lee said that the other lions would probably stay near their wounded fellows or would come back to look for them. We traversed the place where we had

The Wounded Lioness.

chased the antelopes and found that the antelope to which Lee had given the finishing bullet had got up and made off; the quagga was there dead enough, lying on his stomach with his fore leg doubled under him, with his neck arched and striped skin, looking a lovely object. We then made a detour so as to approach the lions from the point from which we originally came upon them. Lee's skill in finding his way in this veldt, where one spot looks exactly like another, was simply marvellous. Getting near the place, I put Walden into a tree with instructions not to descend until he heard me whistle, and proceeded, with the dogs ranging about. The place where the first lion which was wounded had gone we gave a wide berth to, and went straight to the place where the second wounded lion lay. When we were within a few yards of the spot we heard very clearly that peculiar growling, grunting, sobbing sound to which I have before alluded. Lee said, "That means you are to come no nearer"; upon which we retreated a little and consulted. The grass was so thick that we could not see the lion. I suggested that we should climb into trees and fire shots into the patch to see what state she was in, and possibly to move her out of it. This plan was adopted, and having tethered the horses away some distance, we approached and ascended two trees which overlooked the particular patch of grass. Here Myberg's strength and stature served to great advantage, for I stood upon his shoulders and ascended some twenty feet high into the tree. Lee and Myberg ascended another close by, from which, fortunately, they were able to discern the whereabouts of the lion. They fired three shots, and the cessation of the growling, which till then had been continuous and distinct, showed that the bullets had done their work. We then descended and walked into the grass, and found the lion, or rather, the lioness, dead. She had an awful wound in her left shoulder, which was completely shattered by the Martini-Henry bullet used by Lee, and we now saw that she never could have charged us. A bullet through the neck from the tree had finished her off. The behaviour of my pointer dog had been very extraordinary. He had ranged about with much freedom and courage, but whenever he approached the spot where the lioness lay, his tail dropped between his legs and he slunk away. The other two dogs were perfectly useless. What are wanted for a lion are good curs which get near the spot and bark and annoy the lion and make him show himself. We had now to look for the other wounded lioness, and this we proceeded to do with great caution. Unfortunately, we failed to find her anywhere. She must have recovered and made off during the two or three hours of our absence. Then there was nothing to be done but to skin the dead lioness. She was an old lady of great size, with her front teeth much worn away. Her skin was in perfect order. Having got the skin on to the back of one of our horses, we went to look for some of our wounded antelope. After much searching we came across a wounded one who made off, and gave us a short, but exciting chase. A bullet from Lee's rifle and one from my smooth bore brought him down; even then he was not dead, and lay upon the ground with head erect, tossing his horns at the dogs who were barking at him. I gave him two bullets behind the shoulder, Myberg gave him one through the neck, upon which he rose to his feet, swayed about heavily for a few seconds, staggered forward a few paces, and then fell down dead. The tenacity of life of these creatures is inconceivable. It was now dusk, and we had some difficulty in finding our way back to camp, distant about two miles. Here we found Captain Williams, who had hunted and badly wounded three quaggas, but had been unable to secure any of them, owing to the poor quality of his steed. So ended what was to me a very memorable day. I had thought when I came to Africa that I would try and shoot a few nice buck, but I had never bargained to come across such a posse of lions. On reviewing the incidents of the day, I came to the conclusion that all had ended very fortunately, and that I had had an exciting experience such as is known to few, and had escaped unscathed. The following morning, Sunday, July 26th, I started off at daybreak with Lee and "the boys" and six mules to find and bring into camp our wounded game of the day before. The quagga and the roan antelope, which latter had been killed on the previous evening, were soon found close by each other. The former was quickly skinned. Two trees, one of fair size and the other smaller, with forked branches, were then selected, felled, and lashed together. Upon this improvised sledge we bound our dead roan antelope, and to it we attached the six mules. These dragged along their burden, not without difficulty, through the bushy and broken ground, and reached camp about mid-day. I was quite done up, having walked some miles, the sun being very strong. Lee, who had been hunting about all the morning for our other wounded antelopes and quagga, returned in the afternoon with the head and horns of one roan antelope, the entire body and bones of which had been devoured in the night by our friends the lions. I have reason to think that another shooting party in our immediate vicinity picked up two or three head of what we had wounded, as we heard several shots in the course of the morning close to our camp.

It had been arranged with Major Giles that I was to start to rejoin him on the 27th, but the attractions of this spot were so great and the game apparently so numerous that Captain Williams and I determined to stay another day. Off we started at daybreak on Monday, the 27th— Captain Williams, I, Lee, and "the Baboon." I decided that we should keep all together, as lions were evidently about, Captain Williams and "the Baboon" having seen much spoor on Saturday, and I did not choose to run the risk of Captain Williams having such an experience as I had had with no one to help him but "the Baboon," who not only speaks no English, but would probably run away if lions appeared. After riding along for more than a couple of hours in Indian file, Lee discovered a herd of about a dozen quaggas. Captain Williams dismounted and fired at one that was facing him, and away dashed the herd with us in hot pursuit. We soon came up with them, and in a spot where the veldt was fairly open I got a capital shot from my horse's back at a quagga that was galloping along about 100 yards away. To my great delight he dropped to the shot. I found afterwards that the bullet had hit him high on the back, which it broke. We galloped on, keeping the game in view, and had four more shots with uncertain effect. This herd of quaggas bounding through the bush presented a lovely appearance. Soon we halted, as it was no use killing any more of them. It would not have been difficult to kill the whole lot. On going back we found the one Captain Williams had originally fired at lying dead, shot through the chest. The skinning of the two dead quaggas occupied the best part of two hours, when we resumed the chase, determined to kill no more quaggas. We had not gone two hundred yards when we perceived a solitary one grazing. As we watched it it lay down. On our approach with harmless intentions, it bounded off, when we discovered it was badly wounded. We galloped after it, thinking it better to finish it off. The chase lasted ten minutes, during which it received three bullets without dropping. A ball from my Fraser express finally settled him. Proceeding to skin him and to water the horses in a pool hard by, we were startled by the sound of a loud "cooey" from "the Baboon," who had been left about half a mile behind us. Lee exclaimed, "He has seen a lion." We remounted quickly and galloped off to "the Baboon," who told us he had met "a great lion" coming from our direction, who had stopped and looked at him, and after a few seconds had moved away. "The Baboon's" manners and gestures in describing the appearance of the lion were most quaint and grotesque. He said he was immense in size, with an enormous black mane, and added that when he saw him he was so frightened that his hat was lifted off his head by his hair standing on end. We galloped in the direction which "the Baboon" said that the lion had taken, but, though we found and followed his spoor some distance, we never saw him, greatly to the distress of Captain Williams. On our way back to camp Captain Williams and I each secured close to the river two cow waterbucks. We were rather unfortunate on this occasion, as immediately after we had fired a fine bull waterbuck with long horns and an equally good koodoo bull made their appearance from the spruit in which the cows had been feeding, and went off unharmed. On the whole, the day had been a pleasant one, and we had had much sport.

So strongly were we possessed by the charms of "Lion Camp" and its neighbourhood, that we were unable that evening to make up our minds to leave it, and although almost out of provisions, tea, coffee, biscuits, flour, jam, all being consumed, we determined to stay on one more day. A resolution was arrived at not to shoot at koodoo cows or waterbuck cows, or quaggas, but to try only for sable antelope and giraffe, of which latter animal we had seen fresh spoor. Accordingly, on the morning of the 28th, as we were hunting through the bush, several koodoo cows and a fine herd of waterbuck cow were seen, who, as if aware of our policy, gave themselves little trouble to get out of our way, and tried our virtuous resolutions highly. Nothing else did we come across, and we returned to camp at two o'clock, tired and disappointed.

The waterbuck is a handsome animal, nearly as big as the roan antelope, with a broad white stripe running round its quarters and underneath its tail. The cow-waterbuck has no horns. Those of the bull are splendid, nearly equal in beauty to those of the sable antelope. The skin of the waterbuck is greatly prized in this country, being said to exceed in excellence for breeches, boots, gaiters, "rheims," and "vorschlaghts," all other hides. In the afternoon Captain Williams went out hunting with Lee. Sharp rheumatic pains, brought on by bathing on Sunday in the heat of the day, which made riding almost torture, kept me to camp. Our two servants had ridden in the morning to the road to try and obtain from other passing waggons some small supply of provisions. Fortunately they fell in with Messrs. Rylands and Fry, who were hunting near our camp, who generously sent us some flour, some coffee, and some sugar. Captain Williams and Lee returned late, having seen nothing. So we determined to start on the morrow to rejoin Major Giles, a journey of three or four days, expecting to find him on the Lundi River.

CHAPTER XII.

DIFFICULTIES OF TRAVEL ON THE VELDT.

The wealth of Mashonaland—We make a speedy trek and overtake our waggons—Further losses by horse sickness—Stuck fast in Wanetse River—The Sugar Loaf and other miniature mountains—A pestilential spot on the Lundi River banks—A word of warning—Viandt, the Boer ostrich hunter—We reach Fern Spruit—Death of my shooting pony "Charlie"—A veldt fire—A day of discomfort and disaster—Providence Gorge—Description of Fort Victoria—Great loss of horses—Advice to intending emigrants.

THE wealth of Mashonaland may be great even beyond all that wild rumour has asserted, but if it is to be made available for mankind, another route thereto will have to be established than that which I travelled over. Not only does the length of the overland road from the south, some thousand miles or more, present difficulty to the traveller or the merchant, but the character of the country traversed, its geological formation in parts, its climatic and hygienic conditions elsewhere, offer insuperable obstacles to any successful commercial enterprise into this region, conducted from the base of Cape Colony or of the Transvaal. These propositions the mere narrative of my journey will, I think, adequately support. Our camp on the Bubjane River was struck at daybreak on the 29th of July. An arduous trek through the bush back to the road lay before us; many difficult spruits had to be traversed; many long circuits made to avoid impassable, though dry, watercourses; much tree and bush had to be felled. Progress was slow, and though the distance from our camp to the road did not exceed twelve miles, to compass it occupied an entire day. Escaping, by much good fortune, all accident, either to vehicles or animals, we reached the main road, in the evening, where it crosses the Bubjane, at the spot where I had last parted from Major Giles and the main body of the waggons. Short of provisions, having consumed all our bread, biscuit, flour, tea, coffee, and sugar, we had little to eat save the remnants of the game we had shot. Captain Williams and Lee passed the day hunting unsuccessfully, chasing two bull koodoo, but securing neither. Now we had before us a three days' speedy trek, some twenty-five to thirty miles a day, to catch up the remainder of the expedition, which by this time, we reckoned, had reached the Lundi River. Mealies for the horses and mules, beans, milk, and wild honey for ourselves, were obtained by barter from the natives, some bread and coffee were begged from the waggons of Messrs. Rylands, also on a shooting expedition, and all seemed to promise a fortunate termination to our hunting adventures. But I was to commence my experience of one of the pests of African travel, the fatal horse sickness. On the morning of the 30th one of the mules drawing the "spider" was observed to be unable to pull, and to be breathing heavily. Nothing could be done for the animal, no remedies were available, nor, indeed, are any of use; the creature had to be left to die by the roadside. At midday another mule was in the same condition, and was also abandoned, and in the evening a third succumbed and died during the night. The "spider" team being now reduced to three mules, three were taken from the mule waggon, the team of which was reduced to ten mules, too small a number for its heavy load. These losses were depressing; it was impossible to say where they would stop, or in what condition the lapse of a few hours only might find us. Up to this time things had gone very prosperously, not only with me on my journey from Johannesburg to Tuli, but also with Major Giles on his long trek through Bechuanaland. No losses of any importance had been sustained, and I was inclined to think that the horse sickness I had heard so much of was of trifling moment, which could be easily avoided by proper treatment and precaution, and which only embarrassed the ignorant or the careless "trekker." On this point I was destined to be undeceived. The Umsawe river was crossed on this day, a small stream with a fair quantity of flowing water. From here to the Wanetse river is a distance of at least twenty-two miles. Rain had set in the previous evening, damp and cold wind prevailed throughout the day; outspanning, cooking, and eating in the open air ceased to be agreeable, the travel generally was dreary and uncomfortable. Of course, persons who were acquainted with the country and its climate averred that such rain was most unusual. I have travelled many thousands of miles, and visited many places of the earth, and I have never yet arrived at any place where I was not informed that the weather was most unseasonable. The loss of the mules and another wet evening made all dispirited and melancholy. On the morning of the 31st, we reached the Wanetse. This is a river of importance, seventy or eighty yards wide at this time of year, with a strong flowing current. The drift is an awkward one; large boulders obstruct the passage, and rocky slippery banks make it almost impossible for the mules to emerge from the river dragging any load behind them. The "spider" crossed in safety, but with tremendous jolts and shakings. The mule waggon reached the middle and stuck hopelessly fast against a large boulder. The conductor and "the boys" all stripped and plunged into the stream. The whips were freely plied, the mules from the "spider" were added on to the team, a jack was placed under the waggon with great difficulty to raise the wheels, but all to no purpose. The mules struggled, plunged, and tumbled about in the stream and on the rocks in an extraordinary manner, so that it was a wonder either that they were not drowned or that they did not break some of their limbs. Finally, the panting, heaving, fatigued, and dripping creatures were released from their useless toil, and a team of twenty-two strong oxen was borrowed from a transport rider who, with his waggons, was encamped on the opposite bank. These, inspanned and attached, immediately, apparently without effort, extricated the waggon from the river and lightly dragged it up the opposing steep incline. Myberg, our conductor, was somewhat chagrined at this incident, for he was proud of having avoided anything like a stickfast till this day, and he entertained strong opinions as to the superiority of mules over oxen, which were now somewhat shaken.

All this country which had been traversed for some days past is thick bush veldt, studded here and there with rocky "kopjes." These kopjes are of various shapes and height, some of them of great beauty, some fantastic, some almost grotesque in appearance. They rise to an altitude of from 100 to as much as 500 feet. Between the Wanetse and the Lundi, one called the Sugar Loaf, said to attain a height of 800 feet, is a most remarkable object. Generally they are rounded rather than pointed at the summit, and their peculiarity lies in this, that they seem to consist of one immense, massive granite boulder, without discernible crack, fissure, or mark of severance. As a rule they do not rise in ranges, nor are they connected with each other. These miniature mountains, often not without grandeur, add greatly to the attractiveness of the wild woodland scenery, and at evening, when the sunset is brilliant, stand out with a sharpness and an originality which long arrests the eye and excites the imagination. After crossing the Wanetse we overtook the waggons of Mr. Maunde, who had been trekking up from Kimberley since May. He generously supplied us with some fresh excellently-baked white bread. Another set of waggons, with whom I found a friend, gave us some pots of jam, so that our evening supper was, by com-

The "Sugar Loaf" Mountain between the Rivers Wanetse and Lundi.

parison, sumptuous. Living in England, where bread is so cheap, so common, and so wastefully consumed, it is impossible to imagine what a delicious luxury it becomes on the veldt to the

CROSSING THE LUNDI RIVER.

Page 181.

TWO MEMBERS OF THE EXPEDITION CROSSING THE LUNDI RIVER.
Page 161.

traveller who has been for some days without it. I would not have exchanged my loaf of bread this evening for all the delicacies of the Paris Boulevards. Major Giles and the waggons were rejoined on the 1st of August at midday. They were outspanned on the northern bank of the Lundi river, and had lain there for four days. I expect that this delay occasioned by my shooting expedition was in its results somewhat costly to me, and that probably here several of the horses contracted the germs of sickness. The Lundi is a fine river, twice as large as the Wanetse. The stream flows strong and deep, and the water poured into the "spider," rising up to the inside seats. The bottom of the drift is good hard sand, and at this time of year the passage is effected without difficulty. On either side the road descends precipitously to the water, and it is in this portion of the passage that danger lies and accidents often occur. I found Major Giles very impatient to get away from this river. The outspan certainly had a tainted and pestilential appearance. Here for months during the rainy season had lain troops of waggons and of cattle. The camping ground was a mass of dried and fresh dung, smelling disagreeably. Ten yards from our waggons was to be seen a grim array of fourteen graves. Twice as many more, I was told, could be found in the vicinity, and testified to the poisonous fever for which this river has an evil reputation. What had occurred was that many parties of travellers leaving Mashonaland too late in the season last year had arrived at the Lundi to find it a foaming torrent altogether impassable, had been detained there, some for weeks, some even for months, had been without provisions or proper food, and getting low in bodily and mental health, had in many cases succumbed to the malarial fever, for the treatment of which there was neither medical attendance nor medicines. On the morning of my arrival one of our horses, a grey gelding, an excellent animal, well trained to shooting, was taken ill. Quinine, gin, mustard poultices were promptly administered, but the horse died at sunset. Here also one of our oxen strayed and was never recovered. We struck our camp at 4.30 in the afternoon, and trekked eight miles. The evening was saddened by the death of Fly, the grey gelding, the first out of a lot of thirteen horses which had come all the way from Kimberley. Another of the mules in the "spider" team also dying, on the following morning I had to inspan two of the horses. Myberg and Lee took the places of the two men who had driven the "spider" hitherto, and I hoped to proceed with less misfortune. On Sunday, the 2nd August, we traversed for fourteen miles a magnificent country, hilly, well-watered, the bush veldt being more open and park-like than before, dotted with many and various fine trees, covered thickly with sweet grass, good for oxen, with a soil capable of growing every species of agricultural produce. This good country extends from the Lundi to within a few miles of Fort Victoria, a distance of about sixty miles, and seemed incomparably the best part of Mashonaland which I had seen. No finer tract of land for farms could be found in Africa were it not for two fatal disadvantages—(1) the malarial fever, which during the rainy season terribly oppresses human beings; (2) the sickness which at all periods of the year kills from ninety to ninety-five per cent. of horses and mules brought into the country. The opening up of the bush veldt, the cultivation of the soil, and some amount of drainage may overcome the former evil and cause it to disappear, as has been the case in other parts of South Africa; the same causes may diminish the severity of the horse sickness. In the Cape Colony, at Kimberley, and in some parts of the Transvaal, horse sickness, which used to be rife, has almost died out, and I suppose that it is not impossible that science may discover some remedy or some successful mode of treatment which may mitigate the rigour of this malady. Till these changes have extensively occurred I am of opinion that agricultural enterprise in this otherwise beautiful part of Africa would be attended with damage, disaster, and catastrophe.

On the evening of the 2nd August, I met, at our outspan, a Boer, by name Viandt, well known to Lee. He had hunted the ostrich for many successive seasons in Mashonaland, and was acquainted with every hole and corner of the country. He told me he knew places to which no white man of the present day had ever been, where there was much gold and extensive old mining workings. Lee guaranteed his honesty and veracity, so I endeavoured to induce him to accompany us. He was transport-riding, and returning to Tuli with three empty waggons. A bargain was at length struck. He was to turn back and go with me for three months to the gold districts, and his waggons and oxen were to be left at a spot hard by our camp, where there was good veldt, in charge of his "boys" till his return. I went to bed delighted with the arrangement, feeling sure that I would be guided to untold treasure. The morning brought disappointment. The Boer ostrich hunter had been unable to persuade his "boys" to remain in charge of his property, which naturally he could not abandon. He had therefore determined to proceed to Fort Tuli. I ought to have bought his waggons and oxen right out, but this idea did not occur to me till the following day, when it was too late to give effect to it. Possibly the man was a humbug; possibly he knows the position of King Solomon's mines. On the 3rd we only trekked ten miles: both on this and on the preceding morning an ox had been unable to rise and had been abandoned. At the Lundi, two oxen had, while grazing, fallen into deep spruits and broken their necks. Our troop of oxen had fallen below 100. We still passed through an excellent land, in which Myberg and Lee, both high authorities, declared they would wish to possess farms. Another horse, of course one of the best, "Bless" by name, was taken ill this evening. He lived for two days, and we thought we should save him. On Tuesday, the 4th August, we reached Fern Spruit, a lovely spot, where the veldt was very good for oxen, and where there flowed a fresh, sweet stream, between

Passages in the life of one of our boys. In the pantry.

banks clothed with many kinds of fern. During this trek Captain Williams strolled on ahead of the waggon and "spider." Suddenly a fine sable antelope emerged from the bush on to the road, and stood for a minute or two looking at my friend not sixty yards away. Alas! Captain Williams had no arm save a walking-stick, and the antelope passed slowly away unharmed. This kind of vexatious event always occurs when the rifle is left behind. You carry a rifle for days and see nothing. For a moment you leave it in the waggon, and a magnificent chance of sport presents itself. Fern Spruit became hateful to me. We halted here a night and day in order to give "Bless" proper treatment. On the following morning "Ruby," a good horse, was taken ill, and died in less than three hours. Surgeon Rayner made a post-mortem examination of the body. The animal appeared to have been seized with pleurisy, producing a profuse discharge into the bronchial tubes of white foam or froth, a clear yellow serum, which had tried to escape through the windpipe and nostrils, and, by the effort, had produced suffocation. The poor animal had died in two or three minutes with all the struggles and spasms consequent on suffocation. A much harder trial was in store. My shooting-pony "Charlie," to me an invaluable animal, perfectly trained, was taken ill about midday, and was dead in the afternoon. Major Giles and his friends exhausted every effort to save this pony. Every remedy was tried. For a short time sulphur burnt under the nostrils appeared to produce a good effect. During half-an-hour the discharge poured profusely through the nostrils, and if this could have continued, the pony might have survived. So strong was he that three men could scarcely hold him in his efforts to escape from the sulphur fumes. Suddenly the discharge ceased to flow; in a second he fell to the ground

A "Veldt" fire.

and expired almost immediately with desperate struggles, biting the ground with his teeth. I was now very sad, for the pony was a regular pet, and I had grown quite attached to him. We determined at once to quit this malignant locality, and to proceed to Fort Victoria on the high veldt, where we trusted our horses might escape the illness. But this cursed place was nearly fatal to the expedition altogether. The heat of the day was great, the grass long, thick and dry. While we were at luncheon some natives must have set fire to the grass in immediate proximity to, and to windward of, our camp. At the well-known crackling sound of a bush fire we all started to our feet, and beheld that we were surrounded by an advancing mass of high roaring flame, distant from our waggons less than a hundred yards. Not a moment was to be lost, everyone turned out; fortunately we were numerous, forty all told; some seized a coat, some a rug, some a sack, some a branch of a tree hastily torn down, and dashed at the flames to beat down the burning grass and to arrest the progress of the fire. For a minute or two the result seemed doubtful, but by great exertion the fire was overcome and extinguished within a few yards of the camp. At one time I feared that much property would have been consumed and that great and irreparable loss would have been incurred. The natives are in the habit of lighting these fires with perfect carelessness as to who or what may be in the neighbourhood. Afterwards they search the charred ground for dead rats and mice, which they find in quantities and eat. The heat of the sun, of the flames, the violent exercise immediately after luncheon, made two or three of our party very unwell. Altogether this was a hateful and abominable spot, and the day was one of discomfort and disaster. While we were inspanning for the evening trek, the horse "Bless," which we thought to save, was suddenly taken worse, and died in a few minutes. A quarter of an hour more nothing remained of us to mark our stay, save the three dead bodies of our poor horses lying stiff and stark and ghastly on the veldt. I drove on in the "spider" to Fort Victoria through a pass in the hills pompously designated "Providence Gorge." We had all expected to find in this awfully-named passage beetling cliffs, sheer precipices, foaming cataracts, a journey of incident and even peril—in short, all the features of the high Alps or Apennines. But "Providence Gorge" is nothing more than a long valley between two low ranges of hills, gradually and slowly ascending some six or seven hundred feet from Fern Spruit to the plain on which lies Fort Victoria. This plain is of great extent, destitute almost of tree or bush, the horizon broken here and there by isolated hills, at the foot of one of which is situated the fort. The fresh and bracing air of this vast expanse made the change from the low to the high veldt perceptible, and very pleasant. The country stretching away to the north appeared to be fair and attractive, but on closer acquaintance this appearance turned out to be most delusive.

There is nothing very remarkable about Fort Victoria. A small square enclosure, protected by a mound, a ditch, a Maxim gun, surrounded by a cluster of huts built of mud, reeds, and grass, marks the rule of the British South Africa Chartered Company, and the site of what may one day be a populous and prosperous township. The Company at present maintain here a force of sixty-five police, commanded by Captain Turner. I learnt that the natives around had grown somewhat bold in their depredations, and had made off with cattle grazing in the vicinity of the fort. An expedition to one of the Kraals had become necessary, and some appearance of defence on the part of the natives had been made, but when the police advanced at the charge the natives fled in precipitation. The cattle were not recovered, but others were sent in. The health of the force is now good; many cases of fever, however, occurred during the rainy season, and the altitude, 3760 feet above the sea, gives no immunity from malaria. The sickness also during the rains ravages horses and mules. Out of about 600 horses brought here only thirty had survived, and only three had "salted," namely, had the sickness and recovered from it. This horse sickness is a terrific scourge, either for the settler or the traveller. I am surprised that the Cape Government or the Chartered Company do not endeavour to cope seriously with this malady. Scientific investigation of the disease, of the grass and water consumed by the animals, conducted in the locality by experts, assisted by farmers or transport-riders of experience, and carried on patiently for a length of time, would make discoveries of value. The disease is probably acute blood-poisoning caused by some bacillus. This, if discovered, might be cultivated, and inoculation might give protection. It might be well if some authority, British or Colonial, would offer a large reward for the discovery of a remedy or of a successful treatment. The Bechuanaland Border Police have been losing, and are still losing, from 80 to 90 per cent. of their horses. The losses of the Chartered Company have been on a similar scale, and have been equalled by those of private persons. The roadside from Tuli hither is literally strewed with dead bodies of horses and mules. I had been exceptionally fortunate, having only lost four out of thirteen horses, and nine out of twenty mules originally purchased. Many other persons had lost every horse and mule in their possession. Colonel Ferreira, proceeding up country in charge of several horses for the Company, had lost every one. The De Beers Company expedition were in a similar plight. I met Mr. Hugh Romilly outside Fort Victoria returning on foot to Tuli to rejoin his waggons which he had left. He had proceeded up country in a Cape cart drawn by four horses. All had died after crossing the Lundi. Mr. Alfred Beit lost more than half his mules, and on reaching Fort Victoria had to resort to oxen to draw his carriage and light waggons. Many others had sustained similar large and heavy losses. I arrived at Fort Charter with a team of six "unsalted" horses in the "spider," and of twelve mules in the mule waggon, all in good health and condition, and I believe I was the first who had ever brought up so many so far. I would advise any one at home, who might be contemplating an expedition out here next year, to send out an agent some months in advance to the Transvaal to purchase a dozen or more "salted" horses and a score or more of "salted" mules. These would cost from 60*l.* to 80*l.* a horse, and from 20*l.* to 30*l.* a mule. They would sell at a large profit in Mashonaland if good animals. But the agent chosen to purchase must speak Dutch well, and have a perfect acquaintance with the ways and habits of Boer farmers. Otherwise he will not obtain their good animals. The Marico district is the best part of the Transvaal in which to purchase horses and mules.

CHAPTER XIII.

CHARACTER OF THE COUNTRY BETWEEN FORTS VICTORIA AND SALISBURY.

Departure for Fort Salisbury—Our native workmen—Water in the desert—A dreary journey—The country between Fort Victoria and Fort Charter—Where is the "Promised Land"?—We meet Mr. Colquhoun—The garrison of Fort Charter—From Fort Charter to Fort Salisbury—Lions in the way—The Settlement at Fort Salisbury—Signs of civilization—The gold districts of Manica, Mazoe River, and Hartley Hill—Reconnoitring after Game.

MAJOR GILES and the waggons reached Fort Victoria on the morning of the 5th of August. As the year was getting on, and as time was growing short, it was decided that I should go ahead of the waggons to Fort Salisbury, travelling with the "spider" and mule waggon, and arriving at Fort Salisbury, if all went well, ten days or a fortnight before the slowly-travelling ox teams. The morning was occupied in fitting out and provisioning for three weeks this small flying column. My three remaining mules were turned over to the mule waggon, which now possesses a fine team of fourteen. Six horses were inspanned into the "spider," making a most respectable appearance, and I comforted myself with the idea that, reserving a little for a final gallop, I would dash up to Fort Salisbury in imposing style. But I was harassed with doubt and anxiety as to whether this nice fast team would not all die before I reached Fort Charter, and it was with much misgiving that I started on my way at two o'clock in the afternoon of the 6th. Shortly after we had left we were approached by three naked savages, who asked to be allowed to accompany us, and to work for us for one or two months on condition that they should at the end of that time each receive a blanket. We told them that we could spare them no food, as we were all closely rationed. This they thought of no importance, and without any actual agreement being come to, they attached themselves to us, and have followed us till now. They are an amusing and interesting trio—a big savage, a medium-sized one, and a small one. When we arrive at an outspan they set off and collect brushwood for the camp fire, and also take the barrels and buckets down to the neighbouring pool or stream for fresh water, making themselves in this way very useful.

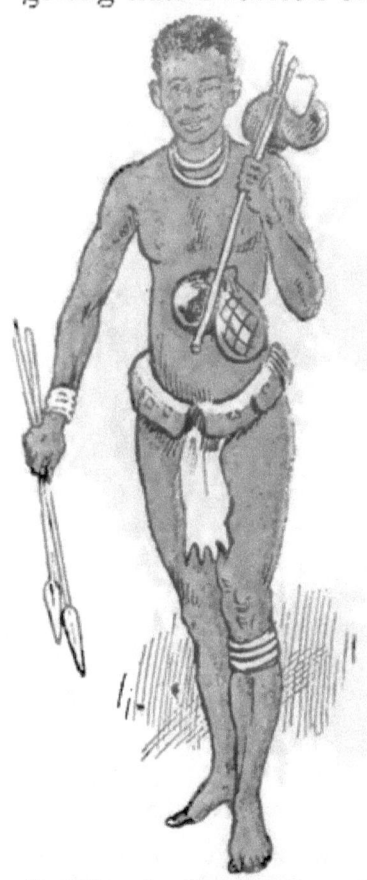

One of our boys (as he appeared with all his household goods).

An Amusing Trio.

Their baggage consists of three dried pumpkin shells, in which they carry a scanty supply of mealies and water. With a little old sacking and half an old blanket found among our odds and ends, they are now fit to appear in Rotten Row. An expression of anxiety and melancholy overcasts their countenances, otherwise intelligent; grave and serious gestures mark their demeanour, not without grace and dignity. They seldom smile, and never laugh. A dreary and tedious journey is the one designated under the heading of this chapter. Five days and a half were occupied in getting over the distance of 125 miles to Fort Charter, though, if the road had been at all decent, less than four days would have sufficed for my horses and mules. The plain which, looked at from a distance at Fort Victoria, seemed so fair, is an endless and most wearisome tract of barren sand, covered with dry, coarse grass, with

One of our boys (in sackcloth, drawing water. Round his neck are two biscuits, gifts mistaken as intended for use as ornament instead of for food).

stunted trees and bushes. No flowing river refreshes this expanse. Water there is in abundance, but of bad quality, lying in stagnant pools, slopping about in marshes and in swamps. Nor is it easy for the traveller to detect from the road the presence of this essential element; the low-lying ground must be carefully searched, or else the much-wanted pool may be passed by. Between eleven and twelve in the morning, and between three and four in the afternoon, the trekker, with mules and horses, must find water if he desires to keep his animals in health and strength. On two days we failed so to find it, with consequent great anxiety and inconvenience. It would be easy for the Chartered Company to cause posts and notices to be erected along the road at places where water is near. Mile-posts or stones along this endless veldt, the maps of which are vague and inaccurate, would again be of inestimable advantage and of facile establishment. If the trekker is to arrange his "scoffs" (that is, journeys from outspan to outspan) well for his animals, he must know where he is and how far he has travelled or has to travel. But this knowledge it is impossible to obtain under present conditions. The natives who are met with have no idea of distance. Asked if a particular place or water is near, they raise the right arm, pointing it straight out to a level with the shoulder; if it is far, the arm is raised higher; if it is very far, then almost perpendicularly over the head. This does not tell one much, and even the traveller of experience may

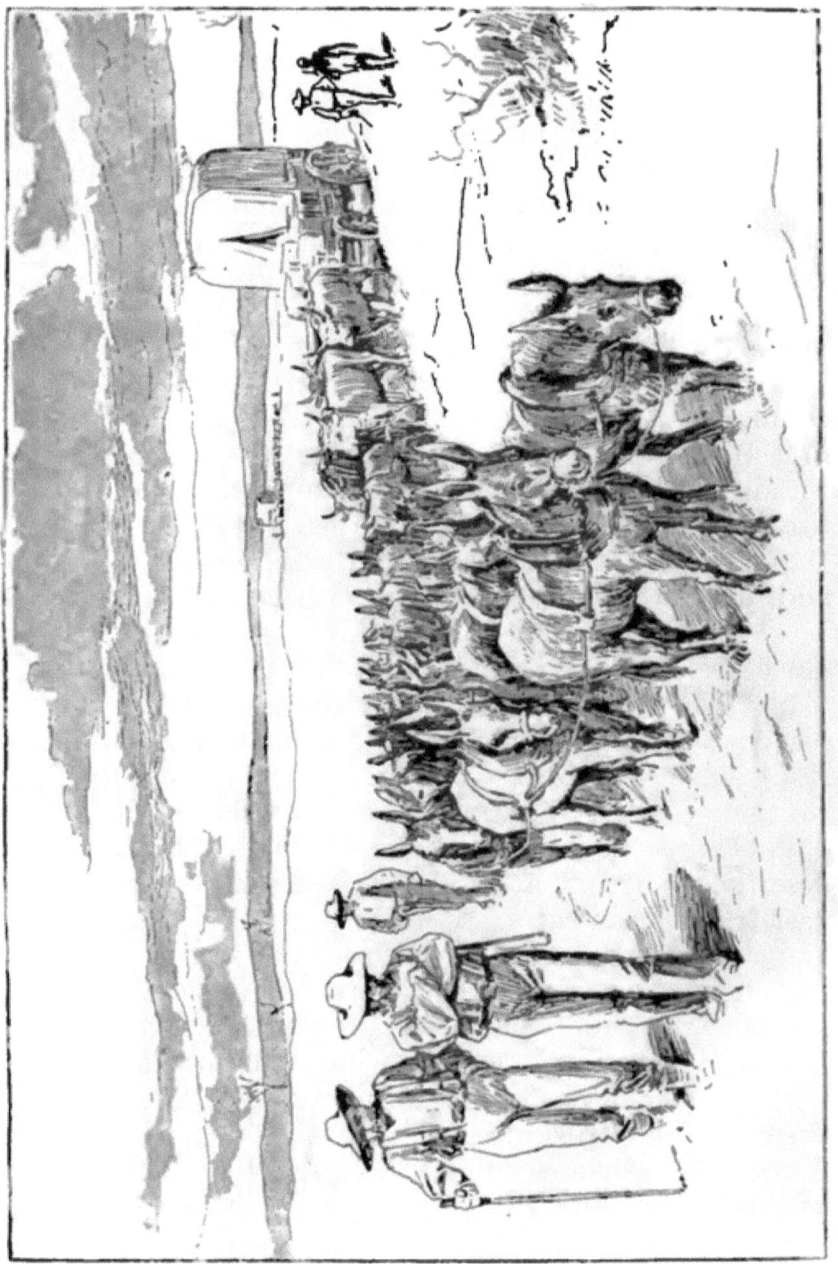

A DREARY ROAD.—THE VIEW FIFTY MILES FROM FORT CHARTER.

Page 197.

often be a dozen miles or so nearer to or further from the place he desires to reach than he is at all aware of. But what makes this stretch of plain from Fort Victoria to Fort Charter so hopeless is the character of the soil and the nature of the grass. Almost the whole way the road lies through heavy sand, in which passing waggons have worn deep ruts, and on which the animals obtain no firm foothold. For long distances the "spider" and mule waggon could only progress at the rate of two miles and a half in the hour; where the incline in the ground is adverse the mules have to stop to get their breath every three or four minutes, and they reach the end of their trek perfectly exhausted. In addition, the whole veldt on either side of the road is what is called "sour veldt" (that is, coarse, hard, dry grass), distasteful to the animals, especially to oxen, perfectly unnourishing. These two bad features of so long a road render travel along it all but impossible. Water is bad and difficult to find, brushwood for fire scarce and hard to obtain. The surrounding scenery is rendered indescribably dreary at this season of the year by vast tracts of burnt grass, presenting a blackened and charred appearance; and day after day the traveller views with displeasure and dismay his animals becoming more lean, more lanky, more fatigued, more weak, less and less able to proceed with the loads behind them, although, to spare them, the daily distances accomplished are brought down to less than six miles. This is the problem of this part of the route: press your animals and they will lie down by dozens, never to rise again; rest them, let them graze, reduce your hours of travel, they derive no benefit from the food or the repose, and the length of your stay in this horrible plain becomes perilously long. This is the part of the road from the south to Mashonaland which is in my opinion absolutely fatal to the route as a route for commerce or for supplies to any considerable population. The vast tract of country between Fort Victoria and Fort Charter is unsuitable and grievous either for man or for domestic beast. Any profitable cultivation of this sandy soil is impossible. In a few spots here and there the natives raise poor crops of mealies. The climate is capricious and variable. One day the heat is so great that it is difficult to support a flannel shirt; the next day a cold wind, driving rain, or thick fog causes you to shiver in a great-coat and muffler. This cold wind and rain is to be expected at this time of the year at least twice in the course of a moon, producing the worst effects on the spans of oxen.

Where, then, I commenced to ask myself, is the much-talked-of fine country of the Mashona? Where is the "promised land" so desperately coveted by the Boers? On the low veldt, where the soil is of extraordinary fertility, fever and horse sickness afflict human beings and exterminate stock; on the high veldt, where neither of these evils extensively prevails, the soil is barren and worthless. I am told that in the neighbourhood, and in the north and north-west of Fort Salisbury, all the conditions are better, that the land is good and the climate fairly healthy, and I trust that this report may turn out to be true, for having now travelled upwards of two hundred miles through Mashonaland, I have, as yet, seen no place suitable for prosperous European settlements. To one person only would this country be attractive, to the sportsman or the hunter. According to native reports, on which reliance can be placed, game abounds. Antelope of all kinds are numerous, sable, wildebeest, hartebeest, eland, ostriches, all can be found and chased, though good galloping horses will be necessary for success, while the presence of many lions offers an exciting variation to the bold and steady shot. I saw from the road and examined through a telescope, two fine herds of hartebeest, but could not chase them, as all the strength of the horses was necessary for the trek. No incident of any kind marked the hours of our travel. We passed many ox waggons brought to a stop owing to the bad road and the bad grass. Two white men, Englishmen, making the journey on foot excited our envy and admiration. All the time they kept up with us, passing with ease in the daytime our struggling horses and mules. A couple of blankets and a small bundle of trading goods to barter for grain with the natives were all their possessions, a bush or rock on the veldt their only sleeping accommodation, but a cheerful heart and a light step bore them gaily on to an unknown goal and unguessed fortunes. We lost two mules on two successive days from sickness, probably contracted in the low country;

the remainder of the mule team and the six horses, well fed with as many mealies as they could eat, kept fairly well and strong, and two days' rest at Fort Charter completely restored them. About five miles south of the fort, on the evening of the 11th August, I met Mr. Colquhoun, the chief civil administrator of Mashonaland, proceeding down

The camp before Fort Charter.

country to meet Mr. Rhodes. Mr. Colquhoun was good enough to stay and converse for a short time. He seemed sanguine as to the future of the country. The gold findings on the Umfuli River, near Hartley Hill, he told me were expected to turn out of great excellence, and he had had favourable reports about the prospecting on the Mazoe River. Fort Salisbury, he added, was becoming quite a township, with a regular street of huts and tents, possessing two auctioneers. We reached Fort Charter on the following morning. The days and nights of the 10th and the 11th had been wet, cold, and foggy; but on the historic 12th the sun shone out brightly, and I thought of the moors,

Summer sleighing in Mashonaland on the high road during the rainy season.

of the grouse, of the Scotch expresses, of friends proceeding northwards. Fort Charter is the counterpart of Fort Victoria. It lies in the centre of a great plain; as far as the eye can range no hill or

eminence can be seen. Within half-a-mile flows a
rivulet containing a fair amount of tolerable water
at all seasons of the year. Both Fort Charter and
Fort Victoria appeared to be miserably weak con-
structions, which a few thousand Matabele would
probably rush with ease, attacking, as is their
habit, in the dark just before daybreak. There is
nothing to stop the rush of the savage foe, save a
ditch from 3 to 4 feet deep, a mound from 10 to
12 feet high from the bottom of the ditch, and
two or three strands of barbed wire stretched on
weak posts. I thought that something in the
nature of *chevaux de frise*—something in the
nature of wire entanglements, would be advan-
tageous and easy of construction, but I was assured
that such ideas were quite wrong and foolish. The
officers of the police evidently disdain the Matabele,
and have perfect confidence in their Martini-
Henry rifles and their Maxim gun. I hope they
are right, but the African savage has often proved
himself to be no contemptible foe, even against arms
of precision. The precaution of sinking a well
either within or close by the fort has been taken
neither here nor at Victoria, nor at Fort Salisbury.
Surely this neglect is imprudent. The garrison
of Fort Charter consisted of a lieutenant and
twelve troopers, of whom ten were sick with
various ailments. The medical arrangements for
the health of the Company's force appeared to be
altogether inadequate. There was no doctor either
at Fort Tuli, where there are upwards of a
hundred men all told, or at Fort Victoria, where

there are nearly seventy, or at Fort Charter. Hospital orderlies have to do duty as doctors. For three months during last rainy season a force of over sixty men at Fort Charter, with many sick, never saw a doctor. Also, for a long time, throughout the country there was a total absence of medicines or medical comforts. Surgeon Rayner, whose brother died of fever at Fort Charter last January while serving in the police, made an exhaustive investigation into this matter. He came to the conclusion that the medical arrangements made by the Company for the occupying force were most unsatisfactory, indeed, scandalously defective, and that even at the time of his inquiry they were open to grave criticism. It is said that the Company will abandon Fort Charter. The usefulness of the position is not obvious. Situated some miles from the main road in a sandy plain, where no one is ever likely to settle, it commands nothing and offers no protection. The altitude of the fort is 4700 feet above the level of the sea; nevertheless, fever in the rainy season is common and serious.

From Fort Charter to Fort Salisbury is a distance of upwards of sixty miles. Leaving the former place on the evening of the 13th August, we reached the Upper Umfuli on the morning of the 15th. The same heavy sandy road, the same wide tracts of burnt grass which impeded our progress to Fort Charter marked the road to the Umfuli. Here two of the mules gave out, and had to be led alongside of the span for the remainder of the journey. Stories of lions and of their audacity in attacking cattle outspanned at night were common along the road. Every waggon met with had had its adventure with these beasts. One trekker, while journeying, had had his span attacked by a lion. An ox was killed, and the remainder broke loose from the yokes, strayed into the bush, into the darkness of the night, and six were never recovered. We took extra precautions in the way of lighting and keeping up several fires all night round our camp, but we never saw, or even heard, a lion. From the Umfuli to the Hunyani River bush veldt is traversed, in parts thick. The road slightly improves in character. Soon after leaving the Umfuli I went on in the "spider," the mule waggon proceeding with its diminished team at a very slow pace. I reached Fort Salisbury on the evening of the 15th August, having accomplished a distance of thirty-five miles between the hours of 6 a.m. and 4 p.m. This was good travelling for this part of the world. About six miles from Fort Salisbury, after emerging from the bush on to the open plain, the traveller passes on his right hand a large and remarkable native kraal. The small conical-roofed hut and store-places of the Mashonas dwelling here are perched on the various jutting and projecting peaks and points of a large rocky kopje. Access to many of these can only be obtained by means of long poles and rough ladders. Fear of the savage Matabele compels the natives thus to dwell. Lobengula has raided, ravaged, and depopulated

BUILDING A "SCHERM" TO KEEP OFF LIONS FROM THE CATTLE ON THE HUNYANI RIVER.

A SKETCH OF THE COUNTRY FROM MATAPI'S KRAAL.

Page 204.

the country. The traveller can procure neither mealies, cattle, meat, eggs, poultry, nor milk. All has been taken or destroyed. In a year or two it is to be hoped that, under the protection of the Chartered Company things will be changed, that the Matabele raids will be occurrences of the past,

Native paintings on rocks at Mateli's kraal.

and that the poor, starving, hunted, timid Mashona will grow his mealies and possess his flocks and herds in peace. For the present, however, the country gives the traveller no assistance whatever in the way of sustenance either for himself or his animals. Situated in a wide and stretching plain, uncovered by bush, Fort Salisbury is not perceived by one approaching from the south until actually reached. The settlement lies at the foot of and around a long kopje about three hundred feet high, thickly clothed with small trees. About half a mile to the eastward are the fort and surrounding huts, and again, half a mile further on, we find the civil lines, where reside Dr. Jamieson, the civil administrator, and the other officials of the Company. Here I outspanned, and was very kindly accommodated with a hut by Sir John Willoughby. In this climate these huts give excellent shelter. Round, about sixteen feet in diameter, with sharply-pitched conical roofs, built of poles and mud, and thatched with grass, they are warm at night and wondrously cool in the heat of the day. They can be erected by the natives in a week at a cost of from 10*l*. to 12*l*. Quite an imposing number of these huts, among which are interspersed waggons, carts, tents, shanties of every conceivable description, compose the settlement of Fort Salisbury, where resided from 500 to 800 persons. The place had a thriving, rising, healthy appearance. The settlers, hard at work, occupied with one business or another from dawn to dusk, wore an expression of contentment and of confidence. A small river flowing through the plain, not a mile distant from any part of the settlement, yields an abundant supply of water. The soil is dry and stony, all moisture either quickly drying up or running quickly off; the altitude is 40 ft.

short of 5000 ft. above the level of the sea, the air fresh and bracing, and these conditions will probably guarantee the good sanitary state of the township and its people, even during the summer and rainy seasons. In the distance surrounding the plain, from which here and there project rocky eminences and huge granite boulders, are ranges of low hills, among which, to the north, rises Mount Hampden, conspicuous and solitary. In a walk round the settlement the next day, I noticed a hotel where was laid out a *table d'hôte* with clean napkins ensconced in glasses on the table, three auctioneers' offices, several stores, the hut of a surgeon-dentist, another of a chemist, a third of a solicitor, and last, but not least among the many signs of civilization, a tolerably smart perambulator. But the necessaries of life, whether of food or raiment, were luxuries at Fort Salisbury, and costly in the extreme. Bread, meat, butter, jam had risen to impossible prices.

My first inquiries were naturally directed towards ascertaining the extent and nature of the gold discoveries. Little could be learnt. I knew almost as much before leaving London on this important matter as I did after a day passed at Fort Salisbury. Three gold districts attracted the prospector :—1. Manica.—Of this district nothing was then precisely known. Colonel Pennefather and Mr. Selous, who had been there for some weeks, were expected to be back at Fort Salisbury in a few days, and we hoped then to learn something definite about this territory, which had been the subject of so much wild rumour and of so much bitter dispute. 2. The gold reefs on the Mazoe River.—These lie about twenty-five to thirty miles distant from Fort Salisbury in a north-easterly direction. A great many prospectors had been at work here, but all accounts of their discoveries and of their value were vague and uncertain; on the whole, however, not encouraging, a yield of from 1 oz. to 30 dwts. per ton being as much as was claimed for the best reefs. Something very much better than this will be required here for a payable mine under present conditions. No deep shafts have yet been sunk, and the depth of the ore is unknown. The regulation 30 ft. hole had been dug on many claims, which of itself gives little or no information. Water had arrested deeper sinkings, and strange to say, among all the mining plant brought into this country, including three or four stamp batteries, there was not to be found one single pump. 3. The Hartley Hill district.—This a gold district, wide and scattered, about thirty or forty miles to the north-westward of Fort Salisbury. On my arrival at Fort Salisbury I found quite a "boom" of claims pegged out on the Umfuli River, flowing near Hartley Hill. Reports of very rich reefs having been discovered, yielding many ounces to the ton, were common, and some excitement prevailed. At the time of writing these pages nothing definite or precise is known, or can be known, about the gold deposits of Mashonaland. There had been no one in the country possessing expert knowledge, on which reliance could be placed, and, even if there had been such persons, no sufficient development work had been effected to enable an opinion of any value to be formed. Many months, probably a year or two, must elapse before any certainty can be arrived at as to whether Mashonaland is a gold-producing country or not. Even if it turns out to be a country possessing gold deposits, the payable character of these depends entirely upon whether cheap and easy access to them can be gained. Nothing can be more uncertain than the character of the Pungwe River route. Accounts are most conflicting, some persons asserting that this route is quite impracticable, others that it is extraordinarily easy. But even if the latter assertion be true, nearly 500 miles of land transit will embarrass the working of the mines near Hartley Hill. So well aware of this serious fact were the officials of the Company that Sir John Willoughby was commissioned to form a special expedition to discover, first, whether a road can be made from the Hartley Hill district to Zumbo on the Zambesi, a distance of about one hundred and sixty miles, and secondly, to ascertain whether the Zambesi is navigable between Zumbo and Tette. In my opinion, at the present time all that can be said of Mashonaland from a mining point of view is that the odds are overwhelmingly against the making of any rapid or large fortune by any individual.

The mule waggon arrived safely at Fort Salisbury late on the evening of the 16th, the mules being harassed and exhausted. Mr. Henry C. Perkins, the mining expert, who accompanied me, was desirous of proceeding at once to the Mazoe River district, whither had repaired a

Mr. Perkins, the mining expert, on the war-path.

few days previously Mr. Alfred Beit and Mr. Rolker, the mining expert sent out by the Chartered Company. I accordingly borrowed a Scotch cart (a light two-wheeled covered waggon) and a span of eight oxen, with which went also two "salted" horses. Thus equipped and provisioned for more than a fortnight, Mr. Perkins, accompanied by Captain Williams, started for the Mazoe on the morning of the 18th August, expecting to be absent at least a week. I had now nothing to do pending the arrival of my waggons from Fort Victoria, which would not occur, at the best, for a fortnight; and hearing good reports of the shooting between the Umfuli and Hunyani Rivers, and having fortunately secured Sir John Willoughby as a companion, I determined on proceeding on another expedition after buck, and after any other wild animal that chance might put in my way. On the morning of the 18th I had a few hours' ride in the company of Hans Lee in the neighbourhood of Fort Salisbury, towards Mount Hampden. We saw nothing but a reitbuck, a steinbuck, and a jackal, but observed spoor of sable antelope. We got no shot, however. The land round Fort Salisbury in this direction appears to be of fairly good quality, well sheltered, well watered, and well wooded. If Fort Salisbury should ever become an important township, farms here might be very profitable. I came across two enterprising persons who had each in attractive spots marked out the regulation area of three thousand acres, and were busily engaged in erecting huts. They seemed confident of success, and were in excellent spirits. One hour before daybreak on the 19th I left Fort Salisbury in the mule waggon for the Hunyani River, expecting Sir John Willoughby to join me later in the day.

P 2

CHAPTER XIV.

SPORT IN MASHONALAND.

Sport in South Africa—Hints to inexperienced sportsmen—Approximate cost of equipment for a six months' hunting expedition—Sir John Willoughby arrives at our camp on the Hunyani River—Hunting the Hartebeest—How to cook venison—A Slough of Despond—Further hunting adventures after antelope—A native hunting party—A cobra in the camp—Method of scaring vultures off dead game—Accident to Major Giles—Scarcity of grain and food in Mashonaland—Return to Fort Salisbury.

To the young Englishman fond of shooting, of riding, of a wild hunter's life, active, vigorous, healthy, and endowed with adequate fortune, those regions of South Africa which extend from the Limpopo to the Hunyani River offer a field for sport not to be equalled in any other part of the world. During the winter time, from May to September, the climate of this region is almost perfect, the risk of fever slight. The air of the veldt is invigorating, the scenery and surroundings attractive and various, the life of the hunter temperate and wholesome. This man coming to these parts of Africa, eager for sport, will experience little, if any, disappointment. Accompanied and guided by some good Dutch hunter, such as Hans Lee, he will see, pursue, probably kill almost every African wild animal, with the exception of the elephant, buffalo, and rhinoceros. These also may be obtained without difficulty, if one is not daunted by the remoteness of the districts near the Zambesi, by the real rough life incident on the absence of waggons and of all beasts of burden, owing to the existence of the Tsetse fly, or by hard walking exercise under the heat of a tropical sun. But in the vast territory I have defined above, the hunter may without difficulty surround and cheer himself with every species of comfort. Waggons drawn by oxen or by mules, the former are preferable, can penetrate to any part of the bush veldt; tents, bedsteads, provisions of all kinds can be carried with ease, and even a young Pall Mall sybarite would acknowledge that there can be provided out here an inconceivable combination of sport and luxury. The soundest sleep at night, the best of appetites for every meal, the clear head, the cool nerve, the muscle and wind as perfect as after an autumn in the Highlands, are pleasures and delights which can be here experienced, and to which many of our London *jeunesse dorée* are almost strangers. All kinds of strange forest sights, all the beauties and many quaint freaks of nature will charm the eye and exercise the mind. Nor is the exciting element of danger by any means altogether absent. The lion and the leopard are beasts to encounter which successfully requires skill, experience, and courage. Snakes of great venom, some of great size, may not infrequently be met with; falls from the horse when galloping wildly through the bush or over the plain, such as even Leicestershire cannot rival, may occur constantly; and should any one imagine that antelope-hunting in Africa is a tame, safe kind of amusement, three or four weeks' experience of it will easily undeceive him. Then the game. Such numbers, such variety, such size, such beauty. Nothing more wildly lovely can be imagined than the sight of a herd of roan antelope, of hartebeest or quagga galloping through the forest; nothing more wildly exciting than the pursuit of such a herd; sighting the game through the trees, sometimes obtaining a fair standing shot within moderate range, then mounting your horse, loading as you gallop along, leaving him to pick his way as best he can among trees, branches, roots, stones, and holes; coming again within one hundred or one hundred and fifty yards, not dismounting, but almost flinging yourself off your horse and firing both barrels as rapidly and as accurately as you may; then on again, over hill, river, and dale, until you and your steed are alike exhausted. These good chases will not occur every day any more than a good fox chase or a good day's salmon fishing comes often in the season. The hunter may ride for miles and for hours through the most sporting, "gamey" kind of country without setting eyes on a living creature; but when they do occur they are periods of excitement every incident of which the memory cannot fail during a lifetime to retain. Then the accompaniments, the framework as it were, of the chase after buck; the early start, the break of day, the brilliant sunrise, the cool morning air, the return to camp, wearied, but pleased and excited, the bath, the evening meal, eaten with an appetite and a zest such as only an African hunter knows, the camp fire, the pipe, the discussion of the day's sport, the hunter's stories and experiences, the plans for the morrow, no thoughts of rain or bad weather oppressing the mind, all this makes a combination and a concentration of human joy which Paradise might with difficulty rival. Nor is this hunting life, when pursued for a few months or from time to time, a useless, a frivolous, or a stupid existence, especially when it is compared with the sort of idle, unprofitable passing of the time experienced from year to year by numbers of young Englishmen of fortune. Nature and all her ways can be observed and studied with advantage, much knowledge of wild animals and of wild men can be acquired by the observant, the intelligent sportsman, languages may be learnt, habitudes and customs noticed and written about, interesting persons are met with, excellent friendships are formed, the mind and the body are seasoned, hardened, developed by travel in a wild country, all its many incidents, its rough and its smooth, its surprises, its difficulties, its adversities and its perils; and I hold this for certain, that in nine cases out of ten a young Englishman who has had six months of African hunting life, will be a 10 lb. better fellow all round than he was before he started.

These reflections occur to me as I sit in the shade of my mule waggon, encamped within a few miles of the Upper Umfuli, on the banks of a small river, passing the evening moments of a bright and warm

African afternoon in writing these pages. I would not be presumptuous enough to write for the trained traveller or the experienced hunter; rather I put down these following ideas for the possible benefit of those who, like myself, without previous experience or knowledge of this kind of life, are fond of travel, of sport of all kinds, who desire to journey and to stay for a while in strange, almost uninhabited, almost unexplored parts of the earth. A six months' hunting expedition out here need not be a very costly business, at least when compared with the cost of London life to many a young man, and measured by the amount of real pleasure and advantage to be derived from either. For the purchase of a couple of good waggons, and of a couple of spans of eighteen or twenty oxen each, of four or five good shooting horses, £1000 would probably suffice, and if these requisites were purchased with care and skill, much of this outlay would be recovered at the termination of the trip. Some £500 for tents and for the paraphernalia of a camp, for personal wants, for luxuries in the way of food and drink; and an expenditure of £100 a month in wages and food for the boys, grooms, and native followers would keep the expedition going. As for personal outfit, little is required, but that little must be of the best quality. Good tanned buckskin breeches, good strong pigskin gaiters, good brown leather-laced walking boots, a dozen flannel shirts, "a couple of Norfolk jackets," an Inverness cape of warm material, three or four large thick rugs, and a Terai hat, are all that can

Equipment for a Hunting Expedition. 217

be required in the way of clothes in this part of the world. But the very greatest care should be taken in ordering and fitting on all these things before leaving London if inconvenience, vexation, and worry is to be avoided. As for armament, I would suggest a couple of double-barrelled express breech-loading rifles, with rebounding locks, ·500 bore, and about a thousand rounds of ammunition. Solid bullets are greatly to be preferred to expanding bullets. The latter, indeed, in my opinion, are not safe to use in a country where at any moment a lion or a leopard may be met with, as they are so extremely uncertain in their effect upon the animal struck by them. I have seen them kill a buck on the spot, breaking up almost the whole of his inside. I have seen them pierce the fore or hind legs of a buck, inflicting a trifling wound, and I have seen them smash up on the surface of the skin, causing a frightful wound in appearance, but no immediate or necessarily fatal injury. Hans Lee, a high authority and fine marksman, will hear of nothing but the solid bullet. In addition to these rifles, a couple of smooth bores for feathered game, with unchoked barrels, so that ball cartridges may be fired from them, half-a-dozen Martini-Henry rifles for the boys of the camp will complete the outfit in respect of weapons of offence. Hatchets, knives, saws, and any tools should be bought in England of the best makers. In the matter of provisions much can be obtained and of good quality at Cape Town, or at Kimberley, or at

Johannesburg. Tea, coffee, bacon, hams, any wine or liqueurs should be brought out from home. I would strongly recommend that the hunter should provide himself with some champagne. After a long day's hunting in a hot sun this wine is the most refreshing and restoring of all alcoholic beverages. So equipped, the fortunate and persevering sportsman will pass many delightful hours and memorable days. Hunting during a period of some weeks or months, he will probably come across giraffe, hippopotami, ostrich, eland, sable antelope, roan antelope, koodoo, wildebeest, hartebeest, waterbuck, quagga, which latter it is almost a sin to shoot, many kinds of small buck, wild pig, hyæna, and jackal. I was hunting too short a time to meet with all these animals, and was not fortunate enough to see either giraffe, hippopotami, ostrich, eland, or wildebeest. But the spoor of all these animals, with the exception of the sea cow, I saw in quantities, often quite fresh.

I left Fort Salisbury at 9 a.m. on the 19th August, and trekked in the mule waggon as far as the Hunyani River, which was reached at 9 a.m. There we outspanned. A bath and breakfast occupied the morning. Sir John Willoughby arrived about two o'clock in the afternoon, mounted on a sturdy and well-bred grey pony, which had been lent him by Dr. Jamieson. Mr. Borrow, of the firm of Johnson, Heaney, and Borrow, had most kindly lent me two excellent shooting horses, both "salted," for myself and Lee. I found that it would not be possible to ride better animals.

At three o'clock we went out hunting down the course of the Hunyani River within about three or four miles of it. The country here is flat, the bush open; wide grassy plains separated by groves and belts of trees succeed each other. After some time Lee descried a solitary hartebeest grazing. Sir John Willoughby stalked this buck, but could get no nearer than two hundred and fifty yards. He fired two barrels, and, as I thought, hit the hartebeest, who circled wildly round and scampered off. I galloped after him as he made for the bush, and pursued him for nearly two miles, always hoping to get within range in some open space. But he always kept a distance between us of three or four hundred yards and stuck carefully to the trees, bush, and high grass, so that it was difficult to keep him in view, and useless to dismount and fire. These hartebeest are despairing animals to chase. They appear to be cantering along slowly, never exerting themselves, but it requires a horse of great galloping power to overtake them. Their endurance is equal to their speed; it is hopeless work to try and ride them down. They have a wild, weird look, and are the least attractive of all the antelope. In size they are similar to the smaller red deer of the East Coast of Scotland. A whole herd of them when chased sometimes get confused, gallop wildly about, stop to look round, and scatter, giving several good shots to the pursuer; but a solitary one rarely stops or stays, he goes right away, straight on end. Finding my horse was getting blown, and that the bush got thicker, I desisted from the chase, firing a parting but useless shot. My companions soon rejoined me, guided by the report of my rifle. We continued our ride without seeing any more game. Close by the camp, returning home, Lee got a shot at and killed a "duiker," a small antelope. This little beast came in useful, as we had no fresh meat, with the exception of a sheep which had been purchased at Fort Salisbury, and which turned out to be such a wretched, poor animal that it was handed over to the natives who accompanied us. These buck, big and little, are all excellent eating. They are never fat like the park deer of England, or the forest stag in Scotland, but their meat when kept for a day or two is tender and good. The tongue, liver, and kidneys are, in truth, delicacies. Nothing can be more plain and simple than the necessary cooking. A good heap of hot ashes, a couple of baking pans, a little fat, either bacon or butter, lots of pepper and salt, a quarter of an hour's patience, and the best dinner which can be eaten awaits the slayer of the African buck. Roan antelope venison I have found to be most meritorious, but Lee informed me that eland is superior, and that giraffe venison far exceeds either. Nor must I omit to mention the marrow bones of the antelope. Again, the cooking is of the simplest character. The thigh bones, stripped of meat, are thrown upon hot ashes and covered with them. In ten minutes they are ready; a hatchet or a stone serves to break the end of the bone, and such marrow is poured out on the plate as no one in London ever dreamed of. A lucky hunter in Africa need never wait for his dinner, and cannot complain of it. Early on the following morning we trekked towards "Beale's Camp," a locality to which we had been directed, and where we were informed there was much game. This spot lies on the Umfuli River, about eighteen miles south of the Hunyani, between that point of the Umfuli which is traversed by the main road to Fort Victoria and that point which is traversed by the road to Hartley Hill. Shortly after starting we got badly stuck in swampy ground. Spades were used freely to extricate the wheels buried over their axles, two horses were inspanned, but to no purpose. There was nothing for it but to "off load," a most tedious and tiring business. Four thousand pounds weight of load was taken off the waggon. We were only eight in number all told, and a lot of time was consumed. Even then, with the waggon thus lightened, it was all the mules and horses could do to drag it out of and across the swamp to firmer ground. Now all the mass of things "off loaded" have to be carried by us some three or four hundred yards and replaced on the waggon. A real bad business this. It was my first experience of a genuine African stickfast. At one moment it seemed as if we might have to remain in this swamp for days, until a team of oxen could be procured, and as if our hunting trip would be brought to a premature and undignified end. After this we proceeded without mishap along a waggon spoor for about ten miles to a shady grove, overhanging a small rocky river, which appeared to offer an attractive position for a camp. Here we outspanned at midday. A kraal was soon constructed of bush and branches for ourselves, and another for the mules and horses. In the afternoon we started out hunting. Again we came upon a solitary hartebeest bull. I got a good shot at this fellow, as he was facing me about one hundred yards off, but did not hit him. Sir John Willoughby and Lee galloped after him, but failed to secure him. Soon after a sable antelope bull was seen a long way off. Getting off my horse, I crawled to an ant-hill, on looking over the summit of which I perceived my friend at least three hundred yards away. There was no getting nearer to him, so, resting my rifle on the ground, I fired. He also was facing me, and offered but a small mark. Alas! my bullet went but a very few inches to the right of him, and he scampered off, hotly pursued by my companions, who got four shots at him. Lee brought him down. He was a fine old bull, with good horns. It was now dusk, and having "gralloched" the antelope and covered him with long grass and branches to preserve him for a time from the vultures and jackals, we returned to our camp. At daybreak we started off again in the same direction as on the previous evening. Soon we came upon a herd of about a dozen hartebeest, and had a right good chase over two miles or more of varied country. Each of us got four or five shots. Lee, as usual, killed one, a cow; Sir John Willoughby and I wounded one apiece. I saw my antelope separate himself from the herd and make off, and galloped after him. He led me a fine dance, and never gave me but one opportunity of getting near him, which I was too slow to take advantage of. After a two miles' gallop I pulled up, wondering what had become of my companions, and where I was. In about half an hour I heard a shot, and going in that direction found Lee anxiously looking for me. In a chase of this kind it is very easy for the inexperienced to lose himself on the veldt. All landmarks get lost; the direction of the wind, the position of the sun, give little assistance when one has been galloping hard for some distance. I found that all our galloping and shooting had only resulted in the death of one hartebeest. It is wonderful and vexatious in the chase to see how close rifle bullets can go without hitting the animal fired at, and also how often the animal may be hit without fatal effect. After this we rode on for two hours without seeing any game, and were getting near our camp about midday when we observed standing in a grove a fine herd of fifteen or twenty roan antelope. These magnificent creatures cantered off, but soon stopped to look round, giving me a capital shot, as I happened to be in front of the others. I fired both barrels, at a distance of some eighty yards, and knocked down two. One immediately rose again and made off. The uncertain expanding bullet had smashed up on the surface, without penetrating. Lee got a shot at this fellow and knocked him over, but he again got up and followed the herd. We now thought that we would get a good chase, for we had followed them hard for a mile, and the buck were getting blown. Unfortunately, they made for a spruit, with high banks and a muddy bottom, and, while we were searching for a place to cross the stream, escaped away out of our sight. Sir John Willoughby wounded one badly as the herd galloped down to the spruit, and on the other side we found a very bloody spoor, which was followed up for some distance fruitlessly. Then, returning to look for the animal which both Lee and I had hit so hard, we found the place where he had fallen, but of the antelope not a sign. Two natives, who had been following us at a distance all day, came up and promised to spoor the wounded beast, and to bring the horns, which were very fine, into camp. This, however, they failed to do. I think if Lee had himself spoored the animal we should certainly soon have got him; but the day was hot, the horses tired, the camp near, and all seemed to make an immediate dinner necessary. On our way to camp, after covering up the first antelope shot, we saw more hartebeest, but had had enough of chasing for that day.

On the following morning we moved our camp ten miles further on towards the locality we were aiming at. We outspanned under a large and lofty magundi tree. This tree has at this season of the year leaves of the most vivid green, contrasting sharply with the prevailing winter hues, and gives a welcome shade. During our morning trek a herd of hartebeest was seen from the waggon, which Sir John Willoughby pursued. After a long chase he succeeded in killing one close to the spot where we outspanned. Now there were in camp four dead antelope, and much "bill-tong" was made. Four Mashona had made a little hut close by our camp, and gladly assisted in cutting up the meat, of which they received an ample supply. Little native hunting parties are frequently met with in this veldt. Two or three in number, with one wretched old musket and two or three charges of ammunition in common, they rarely kill anything themselves, but trust to finding the dead or wounded game of others, or to being fed by some hunting party such as ours. In default of these resources they subsist on caterpillars, which are found in large quantities on the topmost branches of certain trees. Towards evening I went out for a short ride with Lee in the vicinity of the camp. We got no shot, making three unsuccessful attempts to stalk successively a fine old pauw (bustard), an oribi (sort of gazelle), and two hartebeest cows. Fresh eland spoor was seen, which kindled my hopes of getting a chase after this fine antelope on the morrow. Next day, accompanied by "the Baboon," we hunted in the direction of the Umfuli River. It was a morning of misfortune. Both Sir John Willoughby and I got good standing shots at two solitary roan antelope bulls, and both missed without excuse. Lee also chased and fired at, without result, two koodoo cows. In the course of our ride we arrived at "Beale's Camp." This is a cluster of huts, now for the time deserted, situated on the slopes descending to the Umfuli, which here flows through a spacious plain. Here has resided for some months an enterprising sergeant of the police force, who imagines that he has discovered a rich reef, and has pegged out several claims. The soil differs from the surrounding plain, being of a red sandstone, and the output of quartz was distinct. None of us, however, were able to form any opinion as to its auriferous qualities. Agriculture and cattle-grazing in this neighbourhood might be successful. Lee and I again tried our fortune in the afternoon over the open plain extending to the north. We had a good gallop after and several shots at a herd of hartebeest, of which I succeeded in killing one. Lee this afternoon shot very badly. He carried a rifle of mine, a ·577 single-barrel Henry, and missed shot after shot at comparatively easy distances. He was much put out, and declared that the rifle was a bad one and no use. As he had been shooting with it well on previous days, I could not understand how the weapon could suddenly have become worthless. On our way back to camp as evening was setting in, we got good shots at two roan antelope, which hardly troubled to get out of the way. They were perfectly right, as we both missed easy shots, and Lee was more than ever convinced of the badness of the rifle. During our ride I found a nice little stream and grove about four miles from our camp, all round which place was much fresh spoor of game. To this spot I determined to move the camp on the following day. While we were engaged in moving the next morning, Sir John Willoughby hunted. His grey pony played him a nasty trick, galloping off after he had dismounted to shoot at a reit-buck, and Sir John had to come into camp on foot. One of the grooms was sent out in vain to look for the pony, and "the Baboon's" spooring skill had to be called upon to find the lost animal. He found and brought him into the camp late in the afternoon. With the best-trained and most certain horse it is most unwise to neglect the precaution of attaching the bridle to your waist by a string. If all alone on the veldt, the loss of your horse might be attended by the most disagreeable and even serious consequences. While we were outspanning this morning my servant observed a large snake close to the waggon. I quickly got my gun and shot it while it was wriggling off into some bush. When examined Lee pronounced it to be a cobra of considerable size. The bite of this snake is very rapidly fatal to man or beast. The snake measured 4 ft. 6 in. in length, and was in thickness equal to about three fingers. Broad stripes of dull yellow and grey marked the body. Lee said it was a very rare occurrence to meet with this snake. In the afternoon I took the ·577 rifle and had a long ride, but no shot. Sir John Willoughby hunted towards the Umfuli by himself, and shot a roan antelope bull, not returning to camp till after dark. Next day he went off with "the Baboon" to find and bring in the roan antelope, and I rode with Lee in an opposite direction. We proceeded for hours, passing at times through a most lovely land. Groves and bush of every variety, adorned with every hue of summer, autumn, spring, and winter; glades covered with the greenest grass, growing thick and short from the roots of the herbage which had been burned some weeks previously, presented an ideal of forest scenery such as I would have hardly imagined even Nature herself could have composed on this planet. Nothing was wanted save numerous herds of buck. Not a living creature did we see till noon. Then Lee fired at a reit-buck a long way off, which galloped away. The report of the shot brought out of a plantation three roan antelope, which stood looking at us about two hundred yards away. I dismounted, fired, and away they went. Galloping as hard as I could over an open space, I got within a hundred yards, jumped off, and fired again. This was a lucky bullet. It struck one of the antelope in the thigh. This antelope separated himself from the two others and I made after him. He could not get away from me, and I soon got another shot which finished him. Lee galloped after the two others and shot them both. They were all three fine fat cows, with nice heads. We had now hard work for about two hours, cleaning and covering up the three antelopes which lay on the plain more than a mile apart. Long grass and branches had to be cut and fetched from a distance, and before we

DRAWING DEAD GAME HOME ON A SLEIGH MADE FROM THE FORK OF A TREE.

Page 229.

had finished our attentions to the first the vultures had settled by scores on the other two buck. We saved these, however, before much harm had been done. Vultures and jackals will not venture, for hours, to approach dead game well covered up with grass and branches, fearing a trap. It is a good plan to tie a pocket-handkerchief to a stick over the heap and leave it fluttering in the wind. We reached camp about two o'clock in the afternoon, after a tiring but satisfactory morning's work. "The Baboon" was immediately despatched with the mule driver, the two savages, and ten mules to find and bring in the three dead buck. He is unerring in finding dead game on the veldt. Guided only by the vaguest directions, he follows the spoor of the hunter's horse, it may be, for miles, till he arrives at the locality. When only one buck has to be brought in he takes with him two horses or donkeys. The animal is half skinned; half the body, divided lengthways, with the head, is placed on one horse; the other half, with the skin, on the other. On this occasion, when it was a question of three large buck, it was necessary to send a team of mules, and to construct, where the game lay, a timber sledge on which to place the bodies. "The Baboon" and his cortège started off about three in the afternoon, and I hoped that he would have returned to camp before night. He had some five miles to travel to the buck. For once "the Baboon's" skill somewhat failed him. Thinking to make a short cut, he neglected to follow our spoor, and, misled by a white flag, which some prospector probably had set up on a mound, and which he thought was our mark, did not find the bodies of the antelope that evening. He and his party had to pass a disagreeable night on the veldt without food or water. I was somewhat anxious about them, and also feared for the mules. Towards evening a messenger arrived from Fort Salisbury with letters for me and for Sir John Willoughby. He had cleverly followed the windings of our waggon spoor, separating it from old tracks. My letter brought me bad news of my waggons. They had come to a standstill between Fort Victoria and Fort Charter in that hopeless sandy plain which I wrote about in a former chapter, and the oxen, exhausted by the heavy roads and deriving no nourishment on the wide tracts of burnt grass or from the " sour veldt," were, I was informed by Mr. Edgell, totally unable to proceed with their loads. Further, Major Giles had broken his collar-bone while riding a hurdle race at Fort Victoria, and had had to be sent back to the Fort. This was a bad business, but not altogether unexpected by me. When I saw what sort of country it was which had to be traversed, I entertained misgivings as to whether any waggons would ever get across it without much delay and much loss. I had passed, moreover, many troops of waggons utterly unable to proceed. Mr. Edgell begged me to send down to him from Fort Salisbury some fresh spans of oxen. Little did he know what sort of place Fort Salisbury was. Fresh oxen in good condition could not be obtained for love or money.

Every one was wanting them, no one had got them. Even mealies, good supplies of which are essential for horses and mules, if these are to do work, could not be obtained except in scanty quantities, with much difficulty and at great prices. I was asked for a sack of mealies, 200 lb. in weight, 5*l*. 10*s*. There had been no organization in this country during the past season for collecting supplies of grain or food. A little care, forethought, and energy exercised since the close of the rainy season, would have collected, stored, and economized great quantities of forage and of food at the forts and at the various post stations; but nothing had been done, and the Company itself, for the feeding of many animals, depended upon the uncertain and precarious arrival of a waggon now and then bearing a few sacks of grain.

Nothing can be more serious than this state of things in a country where locomotion depends upon the health and strength of your animals, and where the health and strength of your animals depend upon abundant and regular supplies of food. The grass over miles and miles of country had been burnt; nightly conflagrations of grass and bush brilliantly illuminate the horizon in all directions, and day after day the oxen had to travel further and further afield in search even of the " sour veldt" which this country throughout its whole length and breadth alone produces. I did not expect to be able to send much assistance to my belated waggons, but determined to give up shooting and return to Fort Salisbury by easy stages. Slow travelling was forced upon me, as I was without mealies for the mules. Sir John Willoughby left on the morning of the 26th to return to Fort Salisbury, in order to superintend the completion of his preparations for his expedition to Zumbo, on the Zambesi. "The Baboon," with the three roan antelope, did not reach our camp till long after noon. These fine buck lying confusedly on the rough sledge, "the Baboon," the naked savages, and the mules, amid the camp and forest surroundings, suggested a good subject for a sketch. Skinning, quartering, and cutting up the meat into strips for bill-tong now occupied all hands. Order and some degree of cleanliness at length restored, Lee and I rode out for a hunt. We soon started and chased a herd of hartebeest, one of which fell to Lee's rifle. Lee had been shooting to-day and the day before with a rifle lent him by Sir John Willoughby, and regained his usual accuracy of aim. I was unfortunate again in wounding another, which escaped, though we followed his tracks for some distance. Next morning the "boys" were directed to take the mule-waggon back to the camp we had originally occupied, some nine miles from the Hunyani River. Lee and I mounted our horses to make a wide circuit to the same place. A long ride we had, from 6 a.m. to midday, seeing nothing in the way of game save three wild pigs, which we endeavoured, without success, to stalk. On reaching our camp we found that the waggons had arrived without mishap, and in the afternoon again rode out. Five koodoo cows were all we saw, and these we did not care to pursue. The game in this country must have been recently much disturbed by hunting parties. It was a great disappointment not seeing any elands, as I had been assured they were plentiful. The wild, savage-looking, but harmless wildebeest I was most anxious to see and shoot, but could not come across any. The hunter soon tires of the perpetual hartebeest, and even roan antelope pall upon one after a time. At dawn next day we trekked to the Hunyani River, avoiding, by a long detour, the swamp where we had stuck so badly some days before. On this river, at the outspan, I met Captain Coventry, who had been sent by Major Giles on horseback to Fort Salisbury to purchase two spans of fresh oxen. These he had succeeded in obtaining of moderate quality and at a high price, 13*l*. 10*s*. per ox, and was on his way back with them to Fort Charter, where he hoped to find the waggons. He had had and still had before him a long and solitary ride. Also here I found Mr. Alfred Beit on his way from the Mazoe River to Hartley Hill. We had not met since Tuli, and he had much of interest to recount concerning the various troubles, losses, and misfortunes which had beset his travels in this very odd and difficult part of the world.

CHAPTER XV.

THE GOLD DISTRICT OF THE MAZOE RIVER.

In quest of gold—Exploration syndicates—Mashonaland as a field for emigration—The Mazoe gold-fields—Captain Williams's report—Old workings—The "Golden Quarry" mine—Other mines visited in the district—More disappointments.

In the course of the morning I rode on into Fort Salisbury, a distance of fourteen miles, which I covered in an hour and a half. Here I found that Mr. Henry C. Perkins and Captain Williams had returned from their expedition to the Mazoe River, having had a very interesting time and some very rough experiences in respect of food and shelter. They brought no good report of the gold discoveries in the Mazoe district. Many mines had been visited and examined, but nothing very promising seen. The reefs appear to be similar in character, long, thin, and fairly rich (some of them) on the surface, but in all cases, so far as hitherto worked, either "pinching out" to nothing at a depth of from twenty-five to fifty feet, or degenerating into quartz containing little gold. Both the eminent experts, Messrs. Perkins and Rolker, were of opinion that although here and there were reefs of comparatively limited extent and depth, which might yield a small profit to the small individual miner, nothing had yet been discovered, nor did the general formation encourage much hope that there would be discovered in that particular district, any reef of such extent, depth, and quality as would justify the formation of a syndicate or com-

Mr. Perkins.   Mr. Rolker.

Visit to the Mazoe gold-fields.   Experts at work.

pany, and a large expenditure of capital to purchase and to work it. This opinion had become known when I arrived at Fort Salisbury, and some disappointment, and even despondency, prevailed.

So many hopes had been raised, so many castles built on the strength of claims pegged out on promising-looking reefs, of selected specimens carelessly or ignorantly tested, of reports of inexperienced and even of designing prospectors, that there could not but be a sharp reaction. I was of opinion that, at any rate, it was a great thing to know that there was no gold in the district (at least of any importance), and consoled myself with the reflection that in all probability Messrs. Perkins and Rolker had saved the British public some considerable sums of money. What I have seen since I commenced my travels in South Africa has led me to the conclusion that no more unwise or unsafe speculation exists than the investment of money in exploration syndicates. There are many of these at work here, or on their way out, and most of those which have come under my notice have had their money finely wasted, and their business properly mismanaged. Mainly owing to good fortune, good advice, and to the excellent qualities of those who are conducting my expedition, I have some anticipation of escaping from this country without any appreciable loss of capital; but there are several shareholders at home in exploration syndicates out here who will hardly see again a sixpence of their money. It is, however, far too soon to give any opinion as to the possible gold production of Mashonaland. All hopes are now centred in the Hartley Hill district, and on the Manica territory, both of which I hope to be able to visit. Reports from the former district are bright and alluring, and even those prospectors and speculators who acknowledge the failure of the Mazoe gold reefs, declare confidently that they never for a moment placed those reefs on an equality with the reefs of Hartley Hill. Soon we shall be more perfectly informed, for the mining experts proceed immediately to this latter locality. Even if disappointment again awaits us, there is still Manica to fall back upon, of which territory, for reasons which I cannot precisely define, I personally entertain great hopes. Still, the non-discovery of alluvial deposits, the historic certainty of the existence of those deposits in the past, the quantity of old workings, all reaching to a particular depth and then abandoned, do suggest disagreeable doubts as to whether the people of old days have not cleared the country of its gold wealth.

Nor can I as yet escape from the opinion that, as a field for emigration, Mashonaland is a disappointment. The climate, fine in winter, but in very many parts quite unhealthy for Europeans in summer; the torrential rains of January and February, during which all work has to be suspended and roads become impassable; the prevalent maiarial fever, the various animal pestilences, and apparent general absence of rich deep soil, such as distinguishes the Transvaal, seem to offer invincible obstacles to large settlements of white people. Naturally, if great and rich gold discoveries are made, those settlements will come, and nature's obstacles will be mitigated and conquered. But in the absence of such discoveries I cannot yet perceive that Mashonaland has much to offer to, much to attract the British emigrant. Here at Fort Salisbury, and on some of the high veldt, a few might live and thrive, but the want of any large market would prevent the gaining of wealth. I have been hunting over a wide tract of country from four to five thousand feet above the level of the sea, between the Upper Umfuli and Hunyani Rivers, very beautiful, and fairly dry at this time of year, with apparently rich soil. It was, however, impossible not to detect from many signs and indications that during the whole of the rainy season, lasting for three or four months, this wide stretch of country is nothing but a vast swamp, in all likelihood reeking with malaria. It is quite possible that as the work of exploration progresses general conditions, much better in every way, may be observed and noted. For the present, however, it cannot be unwise or wrong to check the formation of hopes too high as to the value of this most recent acquisition to the British Empire, or of plans for its development too large and rapid.

I append some extracts from the report of Captain Williams on his journey to the Mazoe River gold district:—

Report by Captain G. Williams on Certain Mines.

"Early on the morning of August 18th, Mr. Perkins and myself started for the Mazoe Goldfields with six indifferent oxen and a two-wheeled cart containing our provisions, blankets, etc. About midday we reached the Gweebi River, which proved, in spite of its insignificant appearance, a rather formidable obstacle to our dejected-looking team, which seemed quite unable to make the slight effort necessary to pull us out. However, by completely unloading the waggon and digging the ground from under the wheels, we enabled them eventually to drag themselves to the other side, and without further mishap we reached Mount Hampden at nightfall, and were soon afterwards joined by Mr. Borrow (of the firm of Johnson, Heaney, and Borrow). He had followed us in a Cape cart drawn by salted mules, having very kindly consented to take us to the best claims, and generally show us what was most worth examination. Up to this point the country was flat, treeless, and occasionally marshy. Some spoor was to be seen, but no game was caught sight of except a small buck and some wild turkeys, although we met a party of prospectors, some of whom had just shot an eland, while the others were still out in pursuit of some ostrich. We got under weigh next morning by moonlight, about 5 a.m., and outspanned three hours afterwards on the Tatagora River, where Mr. Perkins had an opportunity of looking at some claims on a neighbouring kopje which had recently been purchased by the firm mentioned above. The hill appeared to have been worked on one side to some considerable extent by the old miners. Several shafts have been cleared of their débris and opened up, but little or no vein was to be seen in any except one recently sunk by the present prospector, who had struck a small 'stringer' of quartz, of which Mr. Perkins took a sample and found it to be of a very low grade. These old workings are of a very singular and persistent character throughout the district, consisting for the most part of circular shafts varying in depth from twenty to eighty feet, and not more than thirty to thirty-six inches in diameter, which have been sunk at all sorts of distances apart, in many cases not more than one foot, and in others as much as fifty or a hundred. No outcrop is apparent at the surface, and nothing at the bottom of the shafts would seem to suggest a likelier reason for the stoppage of work than the gradual deterioration in the grade and size of the veins. How these rich spots were originally found, and why the shafts were so irregularly disposed, are questions of which no one has as yet been able to suggest a satisfactory solution. That they were abandoned in haste is extremely improbable, for throughout the whole of this district only two implements have been found left in the bottom of the shafts, in one case a rude stone chisel, or pick, in the other an earthen pot, similar in shape, size, and material to those in use by the natives at the present day. Any attempt to judge of their age must be the merest guess-work, as for the most part they might be anything from twenty to one hundred years, and although in a few cases it is true that trees of some size are to be seen actually growing in the old shafts, they are of those soft-wooded and quick-growing varieties which require but little time for development. The country in the midst of which we now found ourselves was of quite a different character to that previously traversed. From Mount Hampden we had descended some 500 feet into the valley of the Mazoe, and wooded hills and ridges, grassy vleys, and clear running streams surrounded us on every side.

"Some of these hills are of considerable height, rising to as much as 1000 feet above the level of the plain, but only in isolated instances was any outcrop of sedimentary rock visible, the greater portion of them consisting of granite, with but few volcanic intrusions. Round our outspan several native kraals could be seen perched upon the neighbouring crests, and I believe it is not yet clearly understood whether they select these steep and inconvenient homes from the fear of Matabele raids or upon the score of health. I am rather inclined to take the last supposition, as it is said that the Matabele have not as yet penetrated to this part of the country, and it would take a sharp experience to teach the indolent Mashona that the laziest is not also the best course. On our return to the waggon we found it surrounded by these natives, who had brought mealies, milk, and Kaffir corn to barter with. We were sadly in want of mealies for our horses, as we had found great difficulty in obtaining them at Fort Salisbury, but all our attempts at a deal were fruitless. In vain we offered calico or limbo in exchange for their wares, in vain we tried to seduce them with the glittering blue bead or the empty cartridge-case. They were not to be moved. It seems that these fastidious and pampered barbarians have become nice about the colour of their adornments, and will take nothing but red or white limbo and a peculiar sort of bead known as the red-white-eye. So, disappointed, we pushed on to the claims next worthy of attention, and spent an hour or two examining a shallow shaft and narrow vein which was submitted to our inspection. From there to the huts of the Mining Commissioner occupied the rest of the day, and with the evening came Mr. A. Beit and his party, among whom was included Mr. Rolker, the mining engineer of the Chartered Company. On the following day a lovely ride of about seven miles took us all to visit the 'Yellow Jacket' mine, the property of Messrs. Johnson, Heaney, and Borrow, of which Mr. Perkins and Mr. Rolker made a careful examination. This reef extends some 1500 feet in length, and two shafts have been sunk on it. Here no native workings were seen, and the attention of the prospectors was arrested by the outcrop which extended for some distance and gave very rich pannings. A specimen of this ore which was pounded down on the spot for us gave roughly about 60oz. to the ton. Unfortunately the vein decreases lamentably in size and richness as it descends, and samples taken at the bottom give very poor results. This was a great disappointment, as at the top the future seemed very promising and represented a mining venture with which any one might have been deceived. While returning to camp we came upon about twenty baboons playing at the foot of the hill, some of them of great size, but they were too shy to allow us to approach nearer than about 200 yards. Mr. Beit and his party left the next morning for Fort Salisbury, while Mr. Perkins and myself, under the guidance of Mr. Borrow, started on horseback to visit a series of properties extending to a distance of about forty miles from the Mining Commissioner's camp, determining to depend upon what hospitality we might find for shelter and food. We saw the 'Jumbo' and the 'Golden Quarry,' the last-named being again the property of the enterprising firm before alluded to. The former had but little to recommend it as far as present development permitted to judge. The latter seems to be a large burst of quartz very wide on the top, but, like all the rest, rapidly losing grade and thickness as a lower level is reached. A spot of very rich ore was found on the outcrop, and to work this a small three-stamp battery has been erected. We found it busily pounding nearly a ton a day, and getting very fair results, in proof of which a basin was proudly produced containing about fifty ounces of amalgam, and representing, I suppose, the first 'clean up' as yet made in Mashonaland. From there we rode on to the camp of Count de la Panouse, where we were very hospitably received, and passed the night. The following day a ride of thirty miles brought us to another camp of this syndicate, where the same kind reception awaited us; but, unfortunately, no realization of the somewhat glowing accounts we had received as to the valuable prospects of this property. A mass of old workings surround the camp, and two shafts have been sunk some considerable depth, but at present without cutting the vein at all, so nothing remained for us but to depart the next morning on our long but beautiful ride homewards. During the journey we crossed many bright, clear little rivers; but in the opinion of Mr. Perkins there is not sufficient fall nor enough water in them to justify their employment for anything requiring the generation of much power. The general aspect of the country is the same here as was described above, but the timber is small, and of a soft quality, unsuitable generally for large lumber. There is, however, plenty for firewood, small mining props, and so forth. From August 26th to the 28th we examined the property of the Exploration Company Syndicate, which is large and scattered. The reefs throughout presented the same character, and so far as we have seen appear uniformly superficial, extending longitudinally for considerable distances, but 'pinching' out and losing their gold as depth is attained. The shafts which the prospectors have sunk are principally vertical, and as little drifting work has been done a very small portion of one vein is exposed in each case, which of course makes it difficult to form a very conclusive opinion. Some of the reefs carry gold to a fair extent and in Mr. Perkins' opinion it would be possible to make them pay a little profit by small individual enterprise; but neither the extent of the reefs, the quality of the ore, nor the general formation of the country, so far at least as judgment can be formed on what has been seen, could justify the formation of large London companies for their further development."

CHAPTER XVI.

HUNTING THE ANTELOPE ON THE HIGH VELDT.

We start for Hartley Hill—The Mashonas as servants—Marriage in Mashonaland—All alone on the Veldt—Hints to hunters when lost on the Veldt—A Kaffir kraal—Barter with the natives—Dangerously bad shooting—The troubles of trekking—The country between Fort Salisbury and Hartley Hill—Wild flowers and fruit—Unsuccessful chase after ostriches—A fine herd of eland—The bull of the herd falls to my gun.

THE record of my travels continues to be mainly occupied with details of the chase. Saturday, August 29th, and the following Sunday and Monday were busy days with all of us. Messrs. Perkins and Rolker were at work from dawn to dusk panning, assaying, and weighing the gold extracted from the numerous samples of quartz which they had brought with them from the Mazoe district gold-fields; while Captain Williams and I had our time taken up with preparations for our journey to Hartley Hill, such as procuring the indispensable mealies and other kinds of provisions, hiring a fresh span of oxen for the Scotch cart. Two roads lead from Fort Salisbury to Hartley Hill; the lower road, said to be about twenty or twenty-five miles the shorter of the two, follows the main route to the south as far as the Hunyani River, after crossing which it branches off to the west, traversing some swampy and difficult ground. The distance by this road to Hartley Hill is computed at about fifty-three miles. The upper road, which I calculate to be nearly seventy miles in length, proceeds first nearly north-west from Fort Salisbury, afterwards turning to the west and south, and this road also crosses in parts swamps and marshy ground. A new road lying between these two and following higher levels is now in course of construction; when completed, communication between these two important centres ought to be greatly facilitated. It may be mentioned that in the rainy season the two existing roads are said to be altogether impassable either for the lightest vehicles or even for horses alone. Our arrangements were that Messrs. Perkins and Rolker, accompanied by Captain Williams, were to proceed as quickly as possible by the lower road, whereas Hans Lee and myself preferred to follow the upper and longer route, along which we were assured we should find considerable quantities of game. I left Fort Salisbury in the mule waggon on the afternoon of August 31st, having made arrangements during the course of the day with a young and enterprising auctioneer for the sale at Fort Salisbury, in the early days of October, of the whole plant of my expedition. This will be an interesting sale as the first of its kind which has taken place in the country. We trekked about seven miles, encamping for the night in a small grove. One of the three natives whom I mentioned in a former letter as joining us on our departure from Fort Victoria left my service this evening without notice. I learnt afterwards that another of the trio who had remained with Captain Williams also bolted the same evening. These two were brothers, and had evidently made arrangements to depart simultaneously. They were very foolish, for they had been kindly treated, well fed, and their work had been light, and by bolting thus they lost the blankets which they would shortly have received as their wages, which are so dearly prized by the Mashona. However, they carried off with them some clothes which we had bought for them, and a fair quantity of "biltong." All opinions concur as to the utter worthlessness of the Mashona as labourers or as servants. They rarely stay more than a fortnight with any one person, and almost always bolt should any blankets or clothing be given them. One of the trio, "Tiriki" by name, alone remained faithful. I had a conversation with him the first evening out from Fort Salisbury, Hans Lee acting as interpreter. He told me he was not related to the two defaulters, and that he was rather glad they had gone, as they two, being brothers, had conspired to put all the work upon him. I did not personally perceive how their departure would lighten his labours. He also told me that he was very anxious to get married, as, if he were lucky, his wife might have daughters whom he would be able to sell in exchange for goats. It seems that in Mashonaland boys are perfectly worthless articles. I asked him how much it would cost to buy a wife, to which he replied that to buy a very pretty wife required seven things, two goats, two blankets, two spades, and some other article to be specified, according to the taste or fancy of the vendor. Tiriki has remained with me till now, and will, I hope, before I leave Mashonaland, be in a position to acquire the wife he longs for, more especially as he told me he had a girl in his eye who might be bought up before long by some one richer than himself.

At dawn the next morning I rode out with Lee on to the veldt, having instructed our boys to make a short trek, and outspan at the nearest water and shade. We rode for some three hours across some fine open veldt, much of which was well watered, and appeared to be suitable either for grazing or for tillage. A good many farms have been marked out in this neighbourhood, and some actually occupied. Observing a solitary kopje in the distance, we rode towards it, and from the summit searched the surrounding plain with a telescope. In a few minutes we descried a herd of about a dozen hartebeest, and about half a mile further to the north a nice herd of roan antelope, among which was to be perceived a fine old bull. These latter we preferred to chase. As a rule, these large antelope allow the hunter to approach, especially if he is down wind, within a distance of three or four hundred yards, standing looking curiously at him. They then make away at a slow pace; you canter after them quietly, increasing your proximity to them. After going a few hundred yards they generally stand and look round again. Now is the time for a good shot if you have got, as you ought to have done, within two hundred yards of them. After the first shot they make off at a gallop, and the hunter must ride his horse to keep up with them, but it is certainly better, if you desire to kill several of a herd, not to press them too closely, contenting yourself with shooting from time to time at a somewhat longer range, and at the same time not exhausting the powers of your horse. In this way, under the guidance of Lee, I have several times chased good herds of buck for twenty minutes or half an hour, getting many shots. With this particular herd the old bull first offered me a fair chance, but I missed him; he galloped off; with my second barrel I hit hard a big cow with fine horns. Seeing she was badly wounded and could not gallop very far or fast, I made after her, and soon finished her with another bullet. Lee in the meantime had galloped after the herd, and had laid low a nice young bull, which later turned out to be the most excellent eating. Having "done the civil" (i.e. grallocked and covered up with grass and branches) to our two dead buck, we turned our steps towards the road. On our way we came across the herd of hartebeest originally seen from the kopje which had been but slightly disturbed by our chase after the roan antelope. I got a longish shot at one of these, and, noticing that he turned away from the rest of the herd, galloped after him, thinking that I had wounded him. I was soon joined in the pursuit by Lee, who dismounted and fired three shots without effect. I continued to chase the buck, who did not seem to be able to get very far away from me. He held on, however, for a long distance, sometimes being as far away as five or six hundred yards, sometimes allowing me to come much closer, when I dismounted and fired. With my sixth shot I hit him in the haunch, and re-mounting soon had the pleasure of seeing his pace get slower and slower from a canter into a trot, from a trot into a walk, finally sinking on to the ground. I rode up to him and off-saddled my panting and dripping horse, now completely blown by a gallop of upwards of three miles over the most varied country, through swampy ground, groves of trees and bush, and over rocky kopjes. This was the first hartebeest I had managed to ride down and shoot by myself, and I was proportionately delighted. I found myself all alone on the veldt, Lee having for some reason or other discontinued the chase. While I was engaged in opening the buck I was startled by a voice behind me. Looking round, I perceived three natives with the usual amount of clothing and assegai. Not knowing quite what the intentions of these barbarians might be, I immediately, with very dignified and lofty signs, ordered them to complete the disembowelling of the buck, and to cut branches and grass with which to cover it up. This they most meekly did, upon which I graciously permitted them to carry away the entrails. I was now somewhat in a quandary, not knowing where I was, being totally unable to discover the way I had come, and Lee being nowhere in sight. However, I knew the road ran west, and that if I took a southerly course I must soon cut it. It may be well for the inexperienced in the chase on the South African veldt to remember an elementary fact, that at midday in the Southern Hemisphere the sun is always in the north, and that to go due south you have only to keep the sun shining on the back of your neck. One or two other little useful facts may be here set down. If lost at night on the veldt on a bright starry evening, four times and a half the length of the Southern Cross, measured from the summit to the base, in the direction of the base indicates the position of the South Pole. The direction of the wind is apt to be misleading, as it generally follows the sun in the course of the day. It is well for a hunter leaving his waggons on the "trek" to make these latter drag a chain between the rear wheels. When in returning from the chase you reach the road, you can easily discover from the presence or absence of the marks of the chain in the dust, among innumerable other spoor, whether your waggons are before or behind you. Sir Frederick Carrington taught me this simple little dodge, which, fortunately for its efficacy, is hardly at all resorted to. It is very imprudent for any one to go hunting on the veldt without a small supply of biscuits and whisky. Chocolate is an excellent thing to carry, and a box of matches is essential. If the hunter towards the evening finds himself really lost, and is a great distance from his camp, it is much better to realize the fact while some daylight remains, and to make timely preparations for passing the night on the veldt, such as choosing a good tree to which to attach your horse, and collecting an ample stock of firewood to last through the night. If these preparations are postponed too long, darkness supervenes, and the hunter is helpless. Also, if being lost on the veldt you happen to kill a buck, choose your resting-place for the night some distance away from the dead game, which is likely to attract either wolves or lions, in whose vicinity at night it is well not to be. I found my way to the road, and shortly afterwards to the waggons, without much difficulty, and despatched the "Baboon" to search for and bring in the two dead buck, which task he successfully accomplished before dark.

At sundown I was surprised by the arrival of Messrs. Perkins and Rolker, who I thought had gone with Captain Williams along the lower road. They brought the somewhat gloomy intelligence that the span of oxen which had been hired for the Scotch cart had strayed and been lost, owing to the carelessness of our boys, and that Captain Williams had remained behind to try and recover them. Up to this moment nothing had been heard of these lost oxen, although parties were sent in all directions to search for them, and Captain Williams eventually arrived at Hartley Hill with another span, which had been kindly lent him by the officials of the Chartered Company. On the following morning Messrs. Perkins and Rolker proceeded on their way, while I contented myself with a four miles' trek to a Kaffir kraal of some size, picturesquely perched on the peaks of a rocky kopje, similar to the one near Fort Salisbury, which I described in a former chapter. Here, under the ample shade of a large "makoona" tree, I made a comfortable camp. We were soon surrounded by a score of natives, old and young, male and female, who, in exchange for the fresh meat with which we were provided, gave us large supplies of mealies and of Kaffir corn. I tried in vain to purchase a goat, as also milk and eggs, but none of these commodities would they produce. About noon, the sun being very hot, Lee and I rode out on our daily hunt. Soon we saw a solitary hartebeest standing under a tree looking at us, at which I fired at a distance of two hundred and fifty yards. After the shot he moved a few paces to the right, and still remained gazing at us intently. Seeing that he was not disposed to make off, and thinking I had missed him, I sat down on the ground, and, resting my rifle on my knees, took steady aim. This bullet hit him in the chest, and he staggered away a few yards and fell. On going up to him I found that my first bullet had broken the lower jaw. This wound must have stupefied him, and probably accounted for his immobility after my first shot. Going on, we again ascend a kopje to spy the plain. As usual, we perceive hartebeest and roan antelope in different directions, and, as usual, prefer to pursue the latter. These were two cows, which, on being chased for a short way, led us on to a herd of seven other roan antelope.

We had a fine gallop of nearly three miles after this herd, getting many shots. I knocked down the old bull, to which in passing, Lee gave what he thought to be a finishing shot. I made after a cow which seemed to be wounded, and rode her to a standstill. I finished her off with a shot fired from the horse's back, and returned to look for the bull. The old fellow, as soon as he saw me approaching, rose to his feet and staggered away for some distance, and then stood and looked at me. Dismounting within a range of less than 100 yards, and sitting down, resting my rifle on my knees, I fired four bullets at him, three of which missed him clean, the fourth penetrating the head and putting an end to his sufferings. This was dangerously bad shooting if a lion had been the object instead of a roan antelope, but a long gallop over the veldt under a hot sun makes the eye and the hand alike unsteady. On the morning of the 3rd of September, "the Baboon" having been sent away to fetch the antelope killed the day before, Lee and I rode in a westerly direction to another Kaffir kraal. Here the natives were making what was evidently a new installation. I noticed that their huts, with conical roofs, and their small circular store-houses constructed of clay and wattle, were erected with marvellous neatness, and even symmetry. I tried to purchase a cow, but the owner declared he would not part with her unless he received two Martini-Henry rifles with sufficient ammunition. One of the natives offered to guide us back, so we followed him. After going for more than four miles, he led us right on to a splendid herd of about forty sable antelope, among which could be easily perceived a regular old patriarch of a bull with long upstanding black mane and wide arching horns. I had always been longing to come across such a herd, but till to-day had only been successful in meeting with single specimens. Nothing more beautiful can be imagined than the sight of this great lot of antelope careering over the plain from two to three hundred yards ahead of us. For some reason or other I shot terribly badly this morning. In a gallop of upwards of three miles I fired twenty-seven cartridges and only actually killed one, although I wounded another which Lee finished. Lee killed three, among which was the old bull, whom I found, on examination, to possess a pair of horns of surpassing excellence. These five antelope lay dead on the ground over a distance of about two miles. I would have pursued the herd further than I did but that a nasty spruit intervened, in crossing which the antelope gained an immense start, and my horse, becoming utterly blown, was unable to make up the lost ground. My last shot, fired at a distance of more than 400 yards right at the herd, brought down a nice young bull, which Lee managed to gallop after and secure. In the afternoon we trekked five miles towards the ground where the sable antelope lay dead, and on the following morning Lee set out at dawn with four horses to find the old bull and the two cows, while "the Baboon" with ten mules started off to bring in the other cow and bull, which lay at some distance away. "The Baboon" brought in his two animals safe and untouched. Lee was only successful in finding the old bull, the natives having probably carried off entire the two dead cows. In the afternoon numbers of natives came in from a kraal in the vicinity. In the short space of two hours the three big buck had been cut up and bartered away and distributed in return for mealies, Kaffir corn, some milk, and a few eggs. The natives are passionately fond of fresh meat, and will give more in exchange for it than for "limbo," wire, or beads. The two following days we occupied in trekking, as time was getting on and Hartley Hill was still distant. During the trek, Lee saw, chased, and killed a good sable antelope bull. During these two days we had to cross a series of swampy places, and three times we stuck fast hopelessly. The tedious and fatiguing process of off-loading, of carrying the goods for a considerable distance, and of again unloading, had to be constantly gone through, to our inexpressible annoyance and disgust. I think on the whole it is better when the waggon first buries its wheels in the mud and sticks to outspan your team and to attach it to the rear of the waggon and to draw this back, than to attempt to drag it forward, when spades, pickaxes, and even off-loading are apt to become unavailing, and you have to wait till some chance passing waggon gives you the assistance of its team, and extricates you from the morass.

The whole of the country which lies between Fort Salisbury and Hartley Hill consists of wide undulating plains, dotted here and there with kopjes and intersected by groves and by long belts of thick bush. In many parts the veldt is covered with a quantity of green plants of many varieties, and with a profusion of wild flowers in full bloom. The petunia grows wild all over this veldt. In the frequent hollows of the plains water is always to be found, and swampy, marshy ground embarrasses and delays your waggon. On the evening of the 6th we encamped on the banks of the Saroe river. Here Lee observed some comparatively recent traces of hippopotami, and early the next morning he and I set off up the river in search of these creatures. But we were unsuccessful, and saw no hippo. This morning I came across a wild fig tree, with much ripe fruit. This fruit is of a soft rose colour, much smaller than the garden fig at home, quite as sweet, with the seeds inside small and dry. These are often much infested by ants, so that one has to be careful in eating these figs. This curious tree seems to have a habit of producing ripe fruit at any season of the year. The only other incident of the day's ride was the appearance of an immense quantity of baboons, the first that I had happened to see since I had landed in Africa. These were very wild, and fled long before we got at all near them. In the afternoon we trekked fourteen miles to the Zimboe river. Here I found outspanned a small party of three men with donkeys, who were in a great state of excitement at having seen close by the road a large herd of elands, as they said, some fifty in number. Next morning Lee and "the Baboon" found the spoor of these elands, and tried for a couple of hours unsuccessfully to follow it. I then again trekked, but had not proceeded far before I overtook one of my friends of the previous evening, who had been out shooting, and had killed a roan antelope, had seen the elands in the distance, as also some ostrich, neither of which, being on foot, had he been able to pursue. On receiving this intelligence, Lee and I immediately mounted our horses, and, leaving the waggon to trek on to Hartley Hill, rode on towards the spot where the elands were supposed to be. On this day I saw a greater quantity and variety of game than I had seen on any other since I began hunting in Africa. We first sighted some large buck, which we took to be elands; getting near them, they turned out to be a herd of seven fine koodoo bulls. I took a shot at one of them at a distance of over 200 yards, but was not successful, the herd galloping off just as I pulled the trigger. We did not pursue them, as eland was the game we were after. We soon came upon the spoor of the eland, quite fresh, indicating their proximity. While we were following it up through a grove of trees we discerned about half a mile out on the plain five ostriches. It was now a question which to go after, the ostrich or the eland, and after much hesitation and discussion we determined to chase the ostrich. We had a good gallop after these for

s 2

more than a quarter of an hour. They took a circular course, and as we occupied the inside of the circle, both Lee and I obtained half-a-dozen good shots each. Alas! we both shot very badly, not one was brought down, but I expect an ostrich going full tilt is not an easy object to hit. Galloping after the ostrich, my horse, putting his foot in a hole, came heavily to the ground. I did not lose my seat, but lost my rifle, which was thrown some yards away. This incident caused delay, and allowed the ostrich to get too far from us, so we abandoned the chase, chagrined at our bad shooting and at not having secured the cock bird, which was in fine plumage. The appearance, however, of these great birds skimming along over the plain with their somewhat grotesque action had been very pleasant and exciting. While we were dismounted, watching the disappearing ostrich through a telescope, three sable antelope approached us and stood looking, about 400 yards away. They seemed to be aware that we had no intention of molesting them, for they kept near us for some time after we had mounted and were riding along, showing no signs of alarm even when they had our wind. The day was very hot, and I have noticed that at mid-day, when the heat is great, and when there is little wind, the antelope are often singularly tame. Now again, being on an eminence, we spied the plain, the sable antelope also spying us close by. To our joy the elands were made out grazing along the edge of some bush about a mile off. The herd was slowly approached, and was seen to be a large one, numbering nearly thirty, with several young calves. At last I had come across these big creatures about which I had heard so much, which I had hoped for so many days and dreamed of so many nights. The herd looked splendid; conspicuous among them stood the old bull, in appearance almost twice as big as the cows, and very majestic. Off they trotted into the bush as we came within 500 yards; after them we cantered, and were soon close on their heels. The eland is not speedy like the roan antelope, hartebeest, or sable antelope; he hardly ever goes out of a trot, but when he is alarmed this trot keeps a horse at a good hand canter to remain within shooting distance. I went after the old bull, who soon left the herd, and, accompanied by a single cow, took over the plain. In a patch of bush this cow abandoned him, and he trotted along all alone, a great, fine beast. Three times I missed him. My fourth bullet hit him high in the haunch, near the tail, when he was about 150 yards away, as he was crossing a spruit. Then I saw he could go no longer, and rode up slowly within twenty yards of him. I shall never forget the sight of this noble and commanding bull eland looking at me most reproachfully, and from time to time moving away a few paces very slowly. There was nothing ungainly or convulsive about his attitude or action, as is often the case with other wounded buck when the hunter draws near. When dismounting I gave him a bullet behind the shoulder, he moved away a few more paces, and lay down quite gracefully on the ground, sighing. Then turning on his side, stretching out wide apart his fore and hind legs, and again relaxing them, he expired in a position of complete repose. In weight he must have exceeded a thousand pounds; his horns were long, straight, thick at the base, with a spiral twist in them. Lee came up, and we had hard work to cut branches and grass sufficient to cover and hide so large a beast. From twenty to thirty pounds of the meat, taken from the back and breast, we cut off at once, and attached to our saddles; and when mounted must have looked like a perambulating butcher's shop. It was well that we carried off so much meat with us, as a grass fire blazing in the vicinity, impelled by the wind, after our departure enveloped and consumed our fine eland bull, and on the morrow "the Baboon" found but a charred carcase, the horns being the only portion of the remains which the flames had been unable to destroy or spoil. A long ride of ten or twelve miles lay before us to Hartley Hill, which we reached about four o'clock in the afternoon, seeing on our way many buck of various sorts, which we disdained to chase.

CHAPTER XVII.

WEALTH OF MASHONALAND.—DOUBT AND DISAPPOINTMENT.

Hartley Hill—Our party again united—The Tsetse-fly pest—Mr. Perkins joins me in a day's shooting—Surgeon Rayner's adventure with a lion—Contemplating the return journey—Making a clean breast of it—Deceptive appearances—Reefs in the Eiffel district—What is to become of the country?—Mr. Perkins and the leopard.

HARTLEY HILL is a low two-peaked kopje, rising out of a plain covered with thick bush. At the foot of the kopje runs the Zimboe, a fresh stream flowing in a rocky bed, which, within a distance of half a mile, joins the Umfuli. This latter river is here a fine piece of water. It was quite refreshing after so long a travel in a comparatively waterless land to find one's self gazing at the long, broad, deep flats which distinguish the Umfuli as a real river from among such a number of capricious and scantily supplied water-courses. There is little of attraction in Hartley Hill itself. The kopje has an unhealthy, stuffy appearance, and its sanitary character corresponds with its appearance. The soil has been much tainted with numerous "outspans." A veritable plague of common black flies persecutes one from morning till evening. For some reason or other the fresh breezes which daily sweep over the veldt scarcely seem to penetrate to, or in any way relieve the oppressive atmosphere of Hartley Hill. On the higher peak of the kopje Messrs. Johnson, Heaney, and Borrow have, with their usual enterprise, erected huts and store-houses. On the other and lower peak, Mr. Graham, the Mining Commissioner,

The mining settlement at Hartley Hill.

has his offices and abode. Many stories of adventures with lions were current when I arrived at Hartley Hill. On the first night I was there a lion broke into the kraal of the firm mentioned above, situated close to where I was outspanned, killed, and carried off a donkey. This lion on two successive evenings returned to his prey, and shots were fired at him without effect. One of the prospectors in the service of Sir John Willoughby, hearing a noise, went out in the dusk of the morning, and seeing three large animals, fired at them with his rifle and laid them all low. Sunrise revealed to him that he had slain three of his master's donkeys. I found on getting to Hartley Hill that Mr. Alfred Beit, accompanied by Messrs. Perkins and Rolker, had gone into tsetse-fly country to inspect some reefs about the wealth of which rumour had been active, situated in the "Eiffel district," and were not expected back for two or three days. On the following morning, to my great relief, Major Giles turned up with a waggon laden with stores and other things of which I was much in want. He brought reassuring news of my waggons and oxen, which he had left outspanned on the Hunyani river, twelve miles from Fort Salisbury. The difficulties and distresses of the expedition had not in reality been so great as had been represented to me in the letter I received three weeks before, while hunting on the Upper Umfuli. The oxen had now found good veldt, where they were picking up strength and putting on flesh rapidly. Captain Williams arrived in the evening from Fort Salisbury, accompanied by Sir John Willoughby, the latter on his way to the Zambesi. On the following day Surgeon Rayner, Captain Coventry, and Mr. Mackay came into camp. Our party then was again united, with the exception of Mr. Edgell, left in charge of the waggons and oxen on the

Hunyani river. A month and two days had elapsed since we separated at Fort Victoria, and it was pleasant and cheering being once more together, all of us having experienced various troubles and adventures, none of us having suffered from any illness or real misfortune. Time passed rapidly in recounting to each other our different narratives, in making plans and arrangements for future operations. Soon Major Giles returned to Fort Salisbury to prepare for the sale of the outfit. Mr. Mackay, with a prospector of some experience, was despatched to the Umswezi river to examine certain reefs which were reported to be rich. This district is infested by the tsetse-fly; neither horses, mules, nor oxen can be taken there. The miner travels on foot, with donkeys carrying his baggage. Donkeys do not enjoy any real immunity from the effects of the bite of the tsetse-fly, but as they appear to resist the poison for a much longer period than any other animal, and as they are of small value, they are found to be of great use for transport in the "fly" country. A certain time having to be passed while reefs were being examined and reports made, I resolved upon another shooting expedition to the locality where I had seen the elands, the ostriches, and many other buck. Mr. Henry C. Perkins, who had had no good shooting, managed to steal a day from his mining business and came with me. We trekked on to the veldt, some ten miles from Hartley Hill, and went hunting on the following morning. A good many buck were seen and shot at, hartebeest, waterbuck, duiker. Mr. Perkins was successful in securing a fine sable antelope bull, after a regular Highland stalk, and an old ram reitbuck. We also chased a large herd of roan antelope, which led us along some terribly rocky and stony ground, where galloping was almost impossible. We managed, however, to kill the old bull of the herd. I remained in this camp for two days after Mr. Perkins had returned to his mines and his reefs, but got little sport. The buck had been too much chased, and had mostly abandoned the locality. On one morning my hopes were excited by coming across a large outcrop of apparently good-looking quartz. This, however, when panned was found not to contain any gold.

Two days after my return to Hartley Hill, Surgeon Rayner and Hans Lee went out to try and get a buck or two, as the camp was in want of fresh meat. A nobler game awaited them. Two lions were seen stalking a herd of roan antelope. The former were at once pursued, and one of the couple soon had three bullets in his hind quarters. Retreating into some high grass, he afterwards charged and chased his hunters, getting rather too close to the doctor to be quite pleasant. This was his last effort, for he was badly wounded, and a bullet in his head terminated his wicked career. As fine a specimen of a lion as could be seen, he measured twelve feet three inches from the tip of his nose to the tip of his tail, his skin was in perfect condition, his mane bushy and dark coloured. The doctor returned to camp greatly pleased with his exciting and fortunate adventure. The indefatigable Lee rode out again in the evening and killed a sable antelope and a hartebeest, so that we were not deprived of fresh meat on account of the lion.

An expedition in this country is almost entirely dependent for fresh meat on the buck which are killed. The natives will not part with their sheep and goats. While I was at Hartley Hill a trader came in from Buluroyo, with slaughter oxen, bought from Lobengula from the Chartered Company. I procured from him two good sheep and a goat. The fat mutton was found by all to be a real luxury. I also purchased from this man two milch cows. These tiny creatures are with difficulty prevailed upon to yield about three bottles of milk a day between them, keeping the rest for their calves, from which the natives never separate them. However, even this scanty quantity of fresh milk was another luxury which had not been enjoyed for weeks or months, and on which we set great store. While at Hartley Hill the increasing heat of the sun indicated the close of the South African winter. Heavy masses of clouds gathering in the afternoon, a sultry and oppressive air, foretold the near commencement of the early rains and storms. Surgeon Rayner ascertained that at midday the thermometer in the shade marked eighty-five degrees. The nights remain cool and fresh, the mercury ranging from forty-five to fifty degrees. Many signs and appearances, however, continue to tell us that our return journey must soon occupy our thoughts. A troublesome prospect this return journey. Eight thousand miles nearly have to be traversed before we see England again. The choice of route also perplexes; whether to retread the weary and monotonous path to Victoria and Tuli, or whether to attempt to reach the coast *viâ* the Pungwe, braving the "fly," the fever, and the discomfort of being deprived of all wheeled vehicles, furnishes matter for frequent and anxious deliberation. The gold district of Manica has still to be visited, but expectation is lowered and hope no longer glows. For now I arrive at the most unsatisfactory portion of my narrative, and have to make a melancholy and mortifying confession.

In the earlier pages of this book I more than once wrote about the wealth and fertility of Mashonaland as of a fact about which there could be neither doubt nor question. An extraordinary concurrence of opinion on the part of many travellers, confirmed largely by historical record and by the traditions of generations, altogether misled me. But the truth has to be told. Mashonaland, so far as is at present known, and much is known, is neither an Arcadia nor an El Dorado. The discovery that the Mazoe river gold district was a disappointment, and that no expectations of fortune could be derived from it, was borne with comparative equanimity, for all were assured, those who had been resident in the country for some time and those who had recently arrived, that the mineral wealth in the district of Hartley Hill would more than compensate for the deficiencies of Mazoe. It seemed impossible that such a mass of apparently substantiated report and of rumour could turn out to be altogether valueless and misleading. I speedily found out, however, that this was the case. Mr. A. Beit with his party returned from their examination of the much talked about "Eiffel" district much disappointed. These reefs are somewhat typical; a considerable outcrop, much of which when broken up shows a wonderful appearance of visible gold; this, however, when extracted by crushing and panning, is found to be of the finest and thinnest character. It seems to have been deposited in small flakes by water filtering through the cracks and crevices of the quartz; so much so that ore, which at first sight might lead even the experienced to hope that it would yield three or four, or even more, ounces to the ton, actually results in the production of from half-a-dozen to a dozen pennyweights. As with "The Eiffel" reefs so with very many others. Again, where the gold is of a coarser and better quality firmly amalgamated with the quartz itself, then the reef is found either to have no appreciable depth, or else at any appreciable depth to yield little appreciable gold. Hardly an exception to these general characteristics has as yet been discovered. A large amount of rumour had found its way to Hartley Hill as to the richness of the reefs on the Umswezi river. Mr. Mackay returned from an expedition there extending over some days, but reported that he had found nothing of promise, and the specimens he brought back with him, when crushed, gave but poor prospects. Another district some fifty miles from here down the Umfuli river, where is situated the Mammoth river about which much talk had been made (as also the Lo Magundi district), is, I expect, of no better character than those I have already written about. Many prospectors, some working for syndicates, some on their own account, many of Australian and American experience, have now been occupied in these districts for some time. Not one, although they are all sufficiently communicative, appears to be able to claim, or to be desirous even of claiming, that he has discovered anything of value or promise. Day by day I see them abandoning the country with the usual expression that "it is not good enough for them." Mr. Henry C. Perkins tells me that he was never yet in any gold district where so few rich specimens of quartz were brought for inspection. The gold reefs of the country are of an exasperating character. When first seen and cursorily examined the general appearance is promising, and hopes are high; but the more they are developed, and the more work is done upon them, the more unpromising and valueless do they become. Such are the facts as at present known about the auriferous wealth of Hartley Hill and the surrounding district. It is, of course, possible that in course of time some fortunate band of prospectors may light upon a really valuable reef, but no consideration or argument that I know of leads me to the expectation that this will be the case any more than the absence of gold in Middlesex would lead me to the expectation that a gold mine would be discovered in Grosvenor Square. I may, however, qualify the unfavourable opinion of the gold-producing capacity of this district expressed above, by the mention of

At Hartley Hill. Panning for gold at Mr. Borrow's hut.

the fact that I have, in conjunction with Mr. Alfred Beit, purchased half the property in one of the mines belonging to Messrs. Johnson, Heaney, and Borrow. The purchase-money is to be expended on the immediate development of the reef, on which, up to now, little or no work has been done. If the reef is found on examination to go down some hundred feet or so—to be three or four feet wide at that depth, and to yield at that depth the same amount of gold which it yields at the surface, then the mine will be one of some value. But looking to the general character of the other reefs in this district I have little expectation that this will be the case. If these general conclusions of mine are correct, and I fear they may be, the question presents itself, and is found to be almost unanswerable, What is to be done with this country? Agriculture on a large scale, cattle-ranching or sheep-farming, except for the feeding of a large mining population, would be a wild and ruinous enterprise. The climate seems to be altogether adverse to colonization and settlement by small emigrants. Moreover, if this region of Africa so exceptionally favoured in some ways by nature is found to be of little value, how infinitely worthless for all European purposes must be the great district of the Central Lakes, the wide possessions of the East African Company, and the much-vaunted Congo State! Sometimes when thinking of Africa as a whole, of Egypt, Tunis, and Morocco, of the Soudan, and of Abyssinia, of the Congo and of the Zambesi, of the many fruitless attempts made by many nations to discover, conquer, and civilize, of the many hopes which have been raised and dashed, of the many expectations which have been formed and falsified, it occurs to me that there must be upon this great continent some awful curse,

T

some withering blight, and that to delude and to mock at the explorer, the gold-hunter, the merchant, the speculator, and even at ministers and monarchs, is its dark fortune and its desperate fate. It is possible, even probable, that these are views too gloomy, formed and set down as they occur to me under the influence of the disappointment occasioned by the discovery that, as in the Mazoe so in the Hartley Hill district, there are probably no gold reefs of value to be acquired. Manica has yet to be visited, and the character of that country may altogether change the colour of my expectations.

On the 25th September I left Hartley Hill to return to Fort Salisbury, and thence to travel towards Manica. Soon I hoped we should be in a position to know, or at any rate to form a tolerably accurate judgment, as to whether Mashonaland is destined to become a prosperous British colony or to remain until the end of time a barren and desolate African expanse. A curious adventure befell Mr. Henry C. Perkins the other day, in which he had a narrow escape from serious personal injury. He and his friends were examining a reef, along which a trench had been cut. At one part of this trench a narrow shaft had been sunk some six feet in depth, at the bottom of which a small tunnelling had been made. Mr. Perkins was on the point of jumping down the shaft to examine the reef, when it fortunately occurred to him that as the sides were steep he might have some difficulty in getting out again.

It was decided to wait before descending until a rope could be procured. While Mr. Perkins and his friends were conversing on the edge of the shaft a roar and a rush was heard, and out bounded amid the startled party a big leopard, which dashed through their legs and disappeared into the bush. This animal had evidently taken up its abode in the little tunnelling at the bottom of the shaft, and if Mr. Perkins had jumped down, as he intended to do, perfectly unarmed, a terrible conflict would probably have taken place between him and the leopard, in a small confined space, from which escape was impossible, and Mr. Perkins would have been very seriously, if not fatally clawed. Snakes and scorpions are constantly found in these old workings and shafts, and explorations and examinations of mines are not without their own special dangers.

CHAPTER XVIII.

LIFE AT FORT SALISBURY.

Mineral wealth of Mashonaland—Reefs in the Mazoe River Valley—The "Matchless" Mine—Good news from Fort Victoria—A personal statement—Enterprise at Fort Salisbury—A model Ranche—Farms leased by the Chartered Company—An interesting auction—Indignation meeting against the Chartered Company—Horse-racing at Fort Salisbury—Organizing the administration of Mashonaland—Mr. Cecil Rhodes's views of the country.

THE formation of any definite and precise opinion about this country, its resources, and its prospects I found to be a matter of difficulty. It cannot be denied that the high hopes which were entertained by so many and various competent authorities as to the great mineral and agricultural wealth of Mashonaland have not hitherto been justified or nearly justified. This much is probably true: that agriculture, while it might be a profitable enterprise for the feeding of a large resident mining population, for purposes of export could not succeed. The soil, which in no part, so far as I have seen or can learn, is of any considerable depth or richness, which over vast tracks is of the most rocky and stony character, which, over other vast tracks is swampy, requiring difficult and costly drainage, does not promise the cheap and easy production of abundant crops of grain. The great length of the communications with the coast and the many obstacles of one kind and another which embarrass those communications, forbid the export of stock, alive or dead. If Mashonaland, therefore, has to rely for its prosperity upon its agricultural capacity alone, it is a country without a future. There remains the question, Is Mashonaland a good gold country? High professional opinion is certainly inclined to answer this in the negative, and to discourage the outlay of capital. Without doubt numerous reefs, which have been found in certain parts of the county, which have been to some extent developed, and from which fair samples have been taken and most carefully assayed, have turned out to be of little or no value. On the other hand, it may be urged that, as the presence of auriferous quartz all over the country, so far as yet explored, is constant, it is not unreasonable to expect that in certain localities yet to be found the quartz will be sufficiently auriferous to ensure profitable working. Hitherto comparatively little prospecting has been done, and much of what has been done has been perfunctorily and ignorantly conducted. Many parties of *soi-disant* prospectors have been fitted out and maintained by syndicates, whose ideas of their duty appear to be that they are to stick to the main routes, lie under their waggons most of the day smoking or sleeping, shoot an occasional buck, and from time to time offer a blanket to some native who will guide them to an old working, where claims can be pegged out, and possibly the regulation thirty-foot shaft sunk. I have seen that a great quantity of money has been frittered away by parties of this kind, and prospecting such as this cannot be taken into account. Some few honest, intelligent, laborious prospectors there are out here, most of them working for themselves, but as yet these men have been able to examine but a small portion of the country. Mashonaland in area is probably larger than the United Kingdom; it has only been occupied for the space of one year, of which less than six months have been available for exploration and prospecting efforts. Obviously it would be hazardous and premature to assert that, because the first gold discoveries are unsatisfactory, no satisfactory discoveries will be made. Many persons who came out here last year and this year supposed that fortunes would be made with great facility, that gold would be found lying about only waiting to be picked up, and such are retiring from the country discontented, and pronouncing the country to be a delusion and a snare. But nature is not prodigal of her gold. In most cases, long sustained efforts, much patience and perseverance are required to win it from her, and sometimes she conceals it so carefully that only the merest chance or accident leads to its discovery. History, tradition, the narratives of many travellers, strongly support the theory that Mashonaland is rich in gold, and the probabilities are that at some time or other these authorities will be borne out. Another year, at least, of carefully conducted and scientific exploration must elapse before any opinion altogether condemning the mineral resources of Mashonaland could be given with any prudence or justification. In the Mazoe River Valley there are many reefs which, while not large enough or rich enough to justify the erection of extensive and elaborate machinery, would certainly, in the opinion of experts, yield a fair profit to a miner with a small capital, or to a group of such men, working cheaply by their own industry and labour. It is probable that by next year the route to Mashonaland by the Pungwe River may be open and easy, in which case the cost of carriage of small stamp batteries would be enormously diminished. As for the Hartley Hill district, at the present moment I can say nothing in its favour. Most of the reefs there were visited by the experts, and fair samples taken from many parts of them. The assays of these samples show that these reefs contain but little gold, and that they are of small depth and extent. To this there is hardly an exception, and I regret to write that the assay of the samples taken from the "Matchless" mine, in which I am personally interested, are at present much below the mark. This district, however, is only a tiny corner of Mashonaland. Up to the time of my return to Fort Salisbury prospects looked gloomy enough, and disappointment and discouragement were prevalent. An improvement, however, occurred. From the Umswezi River district and from the district of Lo-Magundi, persons upon whose opinion a certain amount of reliance can be placed have sent in promising reports. Not such reports as would justify the assumption that important gold discoveries have been made, but which seem to demand the thorough and elaborate prospecting of those districts. From Fort Victoria more important news arrived. Two or three large and promising reefs were discovered by prospectors known to be experienced, whose good opinion of their discoveries had been confirmed by a high authority. Personally I attributed so much importance to this latest find that I altered my plans for the return journey to the coast. Instead of travelling to Manica, inspecting the gold-fields there, and thence to Beira, it became my intention to proceed at once to Fort Victoria, and to reach Cape Colony by the route through Bechuanaland. This plan prevented me from seeing the Manica district, in many ways so interesting, the rainy season being within measurable distance and the journey long. I the less regretted this for the reason that at that time no very good reports of the Manica district had arrived, nor had any good specimens of quartz been brought in. In the face of unfavourable expert opinion of the uncertainty as to the existence of any important gold-field, I clung to the idea that the country would yet reward its possessors and its earlier settlers. On this opinion, or fancy, as some would call it, I acted. Unable to remain in Mashonaland through the rainy season until next year, I established Captain Williams and Mr. Mackay at Fort Salis-

ON THE OUTSKIRTS OF FORT SALISBURY.

Page 281.

bury for a further period to watch and take what part they could in the development of the country. Moreover, I made arrangements for a prospecting expedition to the Lo-Magundi district. In these pages I am aware I have laid myself open to the reproach of writing much about myself; I advance as an excuse—(1) that the personal proceedings of the traveller must form a considerable part of any narrative of travel; (2) speculation in gold mines is attractive and risky. I imagine that very many persons at home are greatly interested in this country, and may possibly be influenced one way or another by reading accounts given by one actually in the land. I fear by the expression of unwarranted hope to excite speculation which may be attended with loss; I fear by setting forth unfavourable opinions to deter speculation which may be attended with gain. I prefer rather to suggest than to pronounce opinion, to recount one's own personal action, and those, be they few or many, who trouble to peruse this record of travel will attribute as much or as little value as they please to my suggestions of opinion, perhaps slightly increasing the value attributed when action and the general tendency of opinion are found to coincide.

The community settled at Fort Salisbury is remarkable for activity and enterprise. Since my first arrival, now two months ago, I observed a noteworthy increase in the size and a marked improvement in the character of the township. Tents and waggon dwellings had rapidly given place to well-constructed huts, and these latter also were being to a considerable extent supplanted by buildings of brick, of which material a fair quality is being manufactured here. Over three hundred building stands had been taken up from the Chartered Company. A building stand measures 100 ft. by 40 ft.; it is liable to a tax of 1*l.* a month. In the first days of the settlement the company granted away these stands without asking for the payment of any premium. As the demand for them increased, the prudent policy was adopted of putting them up to auction, and about one-third of the total number at this time occupied were sold at prices per stand ranging from 1*l.* to 100*l.* A well-situated building stand commanded a good price. I heard of one such, on which had been erected two single-roomed huts and a shed, being sold for 500*l.* The enclosure belonging to Messrs. Johnson, Heaney, and Borrow is the most important and conspicuous in the settlement. Situated on the northern slope of "the kopje," some ten acres in extent, surrounded by a low but massive dry stone wall, it contains large storehouses, stables, and sheds for cattle, a workshop, and a smithy, and is dotted at one end with rows of wheeled vehicles ranging in size and character from the "buck waggon" to the buggy. Higher up "the kopje," among shady trees, is the dwelling-house, mainly constructed of brick, to which leads a broad and well-kept gravel path. Here also is the commencement of a promising garden, the only one in the settlement. The whole place is maintained in a condition of extreme cleanliness and order, and may truthfully be described as a homestead which would be respectable in England and princely in Ireland. The settlement in this country of the three acute and enterprising partners who compose the firm alluded to above has been a fortunate circumstance for the Chartered Company. Whatever they have done has been well done. Their homesteads at Hartley Hill and at Umtala, in Manica, are similar in scale and character to the

Messrs. Johnson, Heaney, and Borrow's rauche at Fort Salisbury.

one here which I have described. Much mining work has been effected by them on many reefs in the various known gold districts, and all of it has been carried out in the best possible manner. I cannot refrain from the observation that in a new country such as this, where one is compelled at times to notice overmuch apathy, sluggishness, unreasonable discontent, and scandalous waste of money, this firm has set a bright example of active perseverance, of intelligent and economical outlay, which encourages the formation of hopeful views on the future of Mashonaland. Agriculture was not being neglected. One hundred and twenty-three farms, mostly in the neighbourhood, of 3000 acres each in extent, had been applied for and marked out. These are leased by the Chartered Company at a rent of 5*l.* a year on the condition that within three months the tenant shall have commenced a beneficial occupation, which means a certain amount of ploughing and sowing of stock and of building. This completed, the farm is inspected by the Surveyor-General of the Company, surveyed, and the title registered in the Company's books. Of this number of farms about twenty had been taken up by Boers. Last week at Fort Salisbury I found specially interesting, as on four days of the week the surplus stock and stores of the expedition which I had brought into the country were being sold off. Messrs. Hopley and Papenfu, assisted by Mr. Slater, the leading auctioneers here, conducted the sale, the result of which was to me very satisfactory. The total sum realized amounted to 2551*l.* The prices fetched by some of the articles are, perhaps, worthy of mention. Timber, deals, and rafters sold at the rate of 16*s.* 3*d.* a foot, showing the scarcity of and demand for good building material; ten gallons of paraffin oil fetched 20*l.*, two gallons of methylated spirits 5*l.*, sporting Martini-Henry and Winchester rifles went from 10*l.* to 15*l.*, two dozen pint bottles of English ale and stout were sold at 3*s.* 6*d.* a bottle, and immediately afterwards retailed at 6*s.* 6*d.* a bottle, common unsifted Boer meal fetched from 8*l.* to 9*l.* a bag weighing 200 lb., common brown sugar (very common) sold at upwards of 3*s*. a pound; butter at 11*s*. a pound, jam at 4*s*. a pot, dried snook (a fish costing about 2*d*. a pound at the Cape) sold at 8*s*. and 9*s*. a pound, tinned hams fetched 2*l*. a-piece, a bottle of

A restaurant at Fort Salisbury.

eau-de-cologne 20*s*., cotton shirts (price in London 9*s*. 6*d*.) were here secured at 33*s*., a new pair of boots fetched 4*l*., an old shooting jacket 25*s*. This enumeration of prices will show that life at Fort Salisbury was somewhat costly. Eighty oxen sold at about 7*l.* 10*s*. a-head, donkeys at about 3*l.* 10*s*., and five waggons at about 50*l.* a-piece, or half their original cost. Money seemed to be plentiful, and the biddings were sustained with great spirit over four days by a small crowd without coats or waistcoats, and with shirt-sleeves rolled up (the regular Mashonaland morning and evening dress), and enlivened by constant chaff, joking, and general good humour. There was great competition for red white-eyed beads, which the savage fashion of Mashonaland prescribed for native attire. Of these I fortunately possessed a good quantity, and there were none and had been none for some time in the settlement; accordingly they went for 12*s*. a pound, their original cost price at Kimberley being about 6*d*. for the same quantity. "Limbo," the coarsest cotton material, manufactured at about 1½*d*. a yard at home, here sold for upwards of a shilling. During this sale I realized with some regret that a large and well-conducted trading expedition into this country would have been a far more profitable speculation than gold prospecting.

The public life of the young and interesting community of Fort Salisbury had early commenced. Some weeks before my arrival a meeting was summoned for the purpose of considering the past action and the policy of the Chartered Company. The meeting was largely attended, and the proceedings were animated, at times stormy. Discontent had arisen mainly owing to the high cost of living, and to some extent presumably to the non-discovery of rich reefs. This smouldering discontent a few persons considered wise and pro-

NEARING THE END.—THE SALE OF THE SURPLUS STOCK AND STORES OF THE EXPEDITION AT FORT SALISBURY.

Page 280.

fitable to attempt to fan into a flame. Very strong speeches were made in denunciation of the company, and of certain of its chief officials, for that they had not brought into the country sufficient supplies of food. These speeches were adorned with the most highly colloquial expressions and interjections. Their authors forgot that the Chartered Company was not responsible for the feeding of other than its own employés, and that if private individuals embarked on the long journey to Mashonaland with insufficient supplies, they had no one but themselves to blame. The mining experts, Messrs. Perkins and Rolker, were also considered by some of the speakers to be responsible for the poverty of the gold discoveries; Sir John Willoughby, for some equally illogical reason, was sharply censured, and the author of these pages held up to odium on the supposition that he had enjoyed certain special privileges with respect to the importation of alcoholic liquor, unjustly withheld from the general body of settlers. The case for the company was courageously and effectually set out by one of its representatives; contradictions on questions of fact were briskly exchanged, and the lie was freely given by one or the other party. The proceedings of this meeting terminated in an orderly manner with the appointment of a vigilance committee with unlimited and unknown powers. This popular commotion was followed by great tranquility. The "vigilance committee" contented themselves with one interview with Dr. Rutherford Harris, the secretary of the company, at which they experienced much difficulty in sustaining their allegations. Since the interview, this formidably-named body neglected their duties, or were afraid to exercise their vast powers. One or more of the leaders left

The first horse race at Fort Salisbury.

the settlement and went down country, curiously enough, just previous to the arrival of Mr. Rhodes, before whom one would have thought they would have been eager to set forth their grievances, display their authority, and who would certainly have been immediately intimidated into compliance with all their demands. But this great opportunity of gaining glory and power the "vigilance committee" pusillanimously allowed to pass; and the population of the settlement I saw growing, progressing, and even prospering under the despotic and grinding tyranny of less than a dozen policemen, whose military duties kept them all day employed at the Fort.

Horse-racing was inaugurated here, but under circumstances which to me, at least, did not appear to be very promising. I had matched a horse which I had sold a few days previously to beat at even weights, over a distance of five furlongs, a horse belonging to Dr. Rutherford Harris. Mr. Slater, the owner of the horse I had nominated, gave his consent to the match. This horse was three-parts bred, and I knew him to possess a good turn of speed, as more than once I had galloped after buck on him. Dr. R. Harris's horse was a good-looking, thick-set brown cob, pig fat. I had little doubt as to the result of the match. In the afternoon, at four o'clock, three-fourths of the population of Fort Salisbury turned out to see the race. The betting varied from six to four to two to one on my opponent's horse. This somewhat alarmed me. Mr. Giffard, the manager of the Bechuanaland Exploration Company's expedition, was to ride Dr. Harris's horse. Sergeant-Major Montgomery rode for me. To my horror the horse which I had nominated appeared on the ground

U

with drooping head and ears, glassy eyes and tucked-up flanks. The other horse looked blooming. Things were getting very "hot." The riders, having weighed out, were started off, and the unfortunate animal which I had matched to be a flyer, tried in vain to canter for fifty yards, and then relapsed into a slow trot, out of which no efforts of his rider could move him. Dr. Harris's horse cantered past the winning post alone. I believe a good lot of money changed hands over this odd business. If Lord Durham or Mr. James Lowther could pay a flying visit here, horse-racing prospects might improve. Otherwise there are three or four jockeys out of employment in England, to whom I can confidently recommend Mashonaland as a congenial sphere for the exercise of their peculiar talents.

The work of organizing the administration of the country proceeds apace. Magistrates have been appointed for the districts of Manica, Victoria, Fort Salisbury, and Hartley Hill. I believe these gentlemen are invested with all the power and authority in civil and criminal cases which is exercised by a judge of the High Court in Cape Colony. A municipal council will soon be elected for the government of the settlement at Fort Salisbury. Its duties will be to frame and enforce sanitary laws and regulations and to maintain the local highways and streets. Half the building-stand tax and a dog-tax are among the sources of revenue to be assigned to this council. The police force has been reduced during the last two months fifty per cent. It now numbers 330 all told. As each policeman costs the company about 200*l.* a year, a very notable economy has been effected. The present strength of the force is probably still much in excess of what will ultimately be found necessary for the peace and order of the country. Postal communication is very slow, irregular, and badly managed. The mails are despatched from Fort Tuli in two-wheeled waggons, drawn by four oxen. These cover a distance of about four hundred miles to Fort Salisbury, at a rate of some twelve miles a day. The drivers, taken from the police force, are under no supervision, and loiter and dawdle along the road to their heart's content. No fine or censure is inflicted when they arrive behind their time. It was reported that the telegraph wire had been laid into Fort Victoria. This work had been carried out with great energy and at considerable cost. The contracts for extending the telegraph to Fort Salisbury are now being carried out, and probably, in about six months' time, Fort Salisbury will be in telegraphic communication with London.[1] A very large and adequate supply of provisions had either been accumulated here by the company or was well on its way up. No fears of scarcity of food or of high prices during the rainy season need be entertained. Already the prices of all necessaries had considerably fallen from the high level of a month or six weeks ago. Stores had been erected by the company at the various

---

[1] The work was completed in February, 1892.

mining centres, where provisions could be purchased at comparatively moderate cost. The financial resources of the company are respectable. The monthly taxes on the building-stands in the townships of Fort Salisbury, Umtala, and Hartley Hill, and of the farming rents may exceed 3000*l.* in the coming year. A nice amount may also be expected from stamps. Licences, moreover, will be a fruitful source of revenue. A general trading licence costs 10*l.*; a hotel licence for the sale of liquor in retail 100*l.*, a bottle licence costs the same. Only three liquor licences have been granted in Fort Salisbury, and it is to be hoped that the company will curtail, as far as possible, this source of profit. I may mention that till within the last few days whisky and brandy were selling at from 50*s.* to 60*s.* a bottle. Regular hours of opening and closing the liquor shops are effectually enforced, and the sale of liquor to natives or coloured men is prohibited under heavy penalties. Speaking generally of the revenue, it is anticipated, and probably on substantial grounds, that, without taking into account any profit from gold mining, the cost of the administration of the country during the coming year will be more than covered. Thus, as I looked all round on the eve of my departure on my journey south, I thought that I could see much that was bright and smiling in the present condition of Mashonaland. The administration in competent hands. The bulk of the settlers who intend to remain on through the rainy season vigorous, confident, and full of enterprise. The one thing needful for the sure prosperity of the land is the discovery of some rich gold-field, and probably the only requisites for the securing of this auspicious event, if it has not been already attained, are patience and hard work.

Mr. Cecil Rhodes arrived from Manica before my departure south. His arrival, long expected

A PARTY AT THE MESS TABLE, AFTER DINNER.
FORT SALISBURY.

and long delayed, his presence in the capital settlement, the knowledge that he was engaged in mastering all the facts and details of the administration, condition, and development of Mashonaland served to stimulate the action of authority, strengthen general confidence, reanimate men's minds. Mr. Rhodes formed a high opinion of the farming capacity of a large district of high veldt lying between this place and Umtala. No physical difficulties of importance, he reported, need obstruct the construction of a railway from the coast to Manica, the ascent from the low to the high land being gradual and easy. Sportsmen at home may like to know that a prodigious quantity of game, big and little, swarms on either side of the Pungwe River. Possibly after next June, July, and August these buffalo, hippo, rhino, and buck of every kind, now neither wild nor wary, will have been frightened away into remote and inaccessible swamps and thickets; possibly before another year is over the silence of the bush between Manica and the coast will be disturbed by the whistle of the steam-engine, by the axe or the pick of the navvy, rather than by the baying of the hound or the crack of the hunter's rifle.

CAPT "TIM" TYSON
OF KIMBERLEY CLUB FAME,
MR RHODES' COMPANION,
WITH HIS PATENT- A CARRIAGE
LAMP AT EACH CORNER OF
THE TABLE.

FORT SALISBURY.—AT THE DENTIST'S.

CHAPTER XIX.

ON THE ROAD HOME.

Second visit to the mines in the Mazoe Valley—Good-bye to Fort Salisbury—Bad roads—The officials of the Chartered Company—Fort Victoria once more—Climate and weather in Mashonaland—Gold discoveries round Fort Victoria—My faithful savage "Tiriki"—We telegraph home from Fort Victoria—Long's Mine—The Lundi River—Bad roads again—Death of a "salted horse"—The journey to Fort Tuli a record "trek."

BEFORE leaving Fort Salisbury I made, in the company of Mr. Cecil Rhodes, an excursion to the Mazoe Valley. We accomplished the distance of thirty miles to the abode of the Mining Commissioner in the course of the day. The road going north passes at the foot of Mount Hampden about twelve miles from Fort Salisbury. Mount Hampden is an isolated kopje a thousand yards or so in length, and some five hundred feet high. This "celebrated eminence" left behind, the road quits the plain and descends into broken and picturesque country, where hills are covered with tree and bush, putting the traveller in mind of the lowlands of Bavaria. Following the valley of the Mazoe river for some distance, we arrived at the "Alice and Susanna" reefs, situated on the right of the road about one hundred feet up the hillside. Here the quartz reef, which is being worked by a syndicate of the Exploration Company, of London, is found to be of a width of two feet and a half at a depth of fifty feet. Samples taken at this depth were assayed to yield, from the Alice shaft, twenty-one pennyweights to the ton; from the Susanna, thirty-three pennyweights. The reef is probably too narrow to supply any large stamp battery, but in view of the fact that nearly every reef in this part of Mashonaland has "pinched out" or become poor in quality on going down, it was satisfactory and encouraging to come across one reef which at a respectable depth held its own. On the following morning at daybreak Mr. Rhodes, Mr. Borrow, and I, rode eight miles through some very beautiful hill country, to inspect the "Yellow Jacket" reef. This reef is very characteristic of the auriferous deposits in Mashonaland yet discovered. The outcrop is seen to ascend, run along the top of, and descend a long kopje with the utmost regularity, and this, tested along a length of fifteen hundred feet, gave samples of the most promising quality. These were assayed to yield from two ounces up to as much as sixty ounces to the ton. It was as fine a gold mining prospect as could be found. Alas! the sinking of two shafts disclosed the mortifying fact that at a very small depth the quartz became poor, and seriously diminished in quantity. A comparatively large sum of money had been put down for the purchase of this reef, and the disappointment of the investor, who reasonably supposed that he had secured one of the finest gold mines in the world, was as great as can be imagined.

A similar ill-fortune pursued the same party in respect of another quartz deposit in the Mazoe district, by name "The Golden Quarry." Here the actual crushing, by a small three-stamp battery, of twenty tons of ore, gave the excellent result of ninety-five ounces of gold. The "Golden Quarry," however, was soon found to be no reef at all, but only a "blow out," or, in other words, a large bunch of quartz which would be rapidly worked out. I should doubt whether, in the history of gold-mining, two more attractive, more deceiving, more disappointing reefs have ever been found than these two which I have written about. We visited also two other mines, the "Warrigal" and the "Mary Pioneer," which at a depth of thirty feet are of a good width and reported to be of good quality, but of these reefs no assays have yet been made on which reliance can be placed. We returned to Fort Salisbury in the evening, thoroughly fatigued by riding for some hours and by jolting in a Cape cart for more hours on a very hot day, but having accomplished an enjoyable and instructive expedition.

On Tuesday, the 20th of October, in the afternoon, I said good-bye to Fort Salisbury. Two months and a week had elapsed since I arrived there. My recollections of the place will be very pleasant and lasting. They will be recollections of good friends, of new and agreeable acquaintances, of a promising community, of a healthy, bright, and breezy locality, of quickly fleeting hours of amusement, of constant and varying interest.

Though valueless, I cannot refrain from an expression of my earnest wishes for the prosperity of the place and of the country of which it is the centre, and while I cannot expect, I rashly allow myself to hope, that by these writings I may rouse in the minds of people at home some amount of active and abiding sympathy for the fortunes of this infant British settlement. My party on leaving was a small one compared with that with which I arrived. Mr. Henry C. Perkins and Surgeon Rayner had preceded me by some days on the road south, travelling in the "spider" with a team of eight horses. In the company of Major Giles and of Mr. Borrow, I travelled in a large covered van or coach on springs, which had been expressly constructed by that firm for passenger traffic along the Pungwe route to Massikessi. This coach, by extraordinary efforts and at a great sacrifice of mules and oxen, had been brought along the route from Beira to Fort Salisbury; but the experiment had convinced Messrs. Johnson and Co. that passenger traffic along the Pungwe river can only be carried on with the aid of steam, and that the tsetse-fly and the many poisonous grasses and herbs which infest the low country are rapidly fatal to oxen, mules, and horses. Consequently they were glad to sell this coach, which just suited me for my long trek of nine hundred miles down through Mashonaland and Bechuanaland. We found that, in this vehicle drawn by twelve mules, we could cover a distance of thirty miles a day, without at all overtasking the strength of the team. The coach held eight

THE OFFICIALS OF THE CHARTERED COMPANY. 299

persons, including drivers, servants, and about three thousand pounds weight of baggage and provisions. Between Fort Charter and Fort Victoria the road is in a shocking condition, much worse than was the case when I travelled up. The heavy sand, which extends for scores of weary miles, had been terribly cut into by the passage of numerous waggons, and progress over this was hopelessly slow. Where the soil was harder, the protruding stumps of felled trees, huge boulders of rock, and ant-heaps were a constant source of danger to a vehicle on springs. It is certainly a great disgrace to the administration of the country that no efforts have been made by it to put this important highway in decent order. The sand, it is true, is incurable, but nothing would be easier than to remove the stumps and rocks and level the ant-heaps. The presence of these results in an immense and unnecessary wear and tear of waggons, and of injury and of loss of draft animals. The officials of the Chartered Company had ready to their hands, in their police, a force well qualified to make and repair the roads. But this force has, since the occupation of the country last year, been maintained in a condition of complete and utter idleness. The men are not even made to keep the forts and the military lines decently clean. The works which have been constructed by them, whether of fortification or of dwelling, are pitiable, showing neither design, skill, nor solidity. The force was offered tracts of ground round the huts for gardens, but these they have neglected even to mark out.

They have no military drill or training, no shooting instruction. Riding post employs a small number of police, but, with this exception, while I was in the country, I was wondering what this most costly force had done, what it was doing, or what it was going to do. It is true that some thirty of the company's police rendered a great service in routing the Portuguese near Massikessi, but the spasmodic energy of a few does not excuse the normal sluggishness and uselessness of the many. Formed of much the same material as the Bechuanaland Border Police, this force in proper hands would have been actively and beneficially employed on public works all over Mashonaland, but I am constrained to remark that the contrast between the police force of the Chartered Company and the Bechuanaland Border Police is startling and deplorable, the latter being as smart, as efficient, and as thoroughly to be depended upon as the former is the reverse. The company have wisely reduced their police from a strength of upwards of six hundred to one of about three hundred men. If they persevere in this policy and abolish the whole force, their financial resources will be largely added to, and no one in Mashonaland one whit the worse.[1]

We reached Fort Victoria after many narrow escapes from smash and overset, at midday on the 26th October. The weather had become very unsettled. On one night the horizon all round was

[1] The police force has recently been almost entirely abolished.

loaded with thunder clouds. The flashes of lightning were scarcely even intermittent, so numerous, constant, and dazzling were they, and the thunder at times appalling. Fortunately for us, the place of our encampment was not within the radius of the storm. I found the climate and weather of Mashonaland from the middle of July to the middle of September to be almost perfect. Two or three rainy days were experienced in August, but as a rule the weather resembled fine, warm summer weather in England. The nights were cool and refreshing, the morning and evening air delightful. After the middle of September the midday heat became oppressive and sultry. The thermometer would mark from eighty-five to ninety degrees in the shade. Every afternoon clouds would gradually cover the sky, and somewhere or other in your neighbourhood, if not actually over you, a heavy thunderstorm would come down. These thunderstorms are disagreeable and even alarming. One of them came over the plain of Fort Victoria on the evening of our arrival. Accompanied by but slight rain, the lightning effects were awful. It lasted the best part of four hours, and was followed by two days of raw damp wind and mist. Camp life under such circumstances is far from pleasant. The regular rainy season of Mashonaland does not usually commence till January, but a sort of foretaste of the regular rains is generally experienced for two or three weeks at this time of year, after which the weather settles again for a time. Travellers going south now become anxious as to the size of the rivers and the condition of the drifts. A heavy flood on the Lundi, Tokwe, or Wanetze might cause a delay of many days in the journey. The vicinity of the road to Fort Victoria has been quite deserted by game and by lions; we neither heard these latter brutes at night nor any stories of them.

Our journey to Fort Victoria was without incident. Three score waggons or more, laden with meal and other supplies, were passed on their way up to Fort Salisbury, and there could be no doubt that this year ample provision of food of all kinds for those who remain in the country during the rainy season had been made by the company. The large agricultural expedition conducted by Mr. Van der Byl was met. All seemed in good heart and order, though the sorrow had been experienced of losing two of their number by death. Good reports of the gold discoveries round Fort Victoria abounded along the road, which turned out to be somewhat fallacious. Four reefs we found had been worked upon, two of which had developed some quartz of a rich character. Not enough work had been yet done on these to determine whether they will turn out mines of great value. The locality abounds in massive outcrops of quartz, most of which rather resemble "blow outs" than regular reefs. Comparing this district with others in northern Mashonaland, the quartz here produced is considered to be of a superior quality, and this district has what is now held to be an advantage, that of possessing no old workings. In the earlier days of the occupation the one great object with everyone in the country was to find an old working, as it was supposed that the ancient miners were unable to work at any depth, or to deal with quartz of great hardness. This theory is probably erroneous. The ancient miners in all likelihood knew more about their business than they are credited with knowing,

Tiriki.

and the abandonment by them of reefs where old workings are now found was due less to their want of skill or knowledge than to the fact that they had worked out the best of the ore.

On the morning of the 30th October I continued my journey south. "Tiriki," the faithful savage about whom I wrote in a former chapter, now departed, the kraal of his tribe being near.

He was as good a specimen of the savage as could be met with, quite intelligent, always cheerful, and willing to work. He entered my service stark naked, but at his departure had accumulated an extraordinary varied wardrobe. Every cast-off pair of trousers, drawers, boots, and shoes, every coat and waistcoat thrown aside had been carefully collected by him, and all that he could not actually wear on his own person was accumulated in an old sack; in this were also many other treasures, smashed pipebowls and stems, empty provision tins, exploded cartridge cases, and every imaginable odd and end. In addition he took with him two blankets, two spades, and two golden sovereigns in

As he arrived.

THE ARRIVAL OF THE TELEGRAPH LINE AT FORT VICTORIA.—SENDING A TELEGRAM TO LONDON.

Page 305.

lieu of the goats which I had promised him, but could not procure. These two latter he concealed away in alternate and numerous coverings of bags, cases, and again bags and wrappings. His figure and appearance when he departed were inconceivably grotesque. He is now probably a millionaire in his kraal, has married the girl whom he has long had in his eye, and as years go by he will add to his wealth by selling his daughters, should fortune still attend him and give him female progeny. Mr. Cecil Rhodes arrived at Fort Victoria a week after our party, on the morning of our departure, and almost immediately rode out to the telegraph wire which had that morning only been brought within two miles of the fort. As it was

As he departed.

X

all on my way I joined him. The scene was peculiar and very African. Amid waggons, oxen, mules, and horses, piles of telegraph poles, coils of wire, boxes of insulators, and odds and ends of baggage and provisions could be seen meandering a little green string communicating with the waggon, which it entered, the elevated wire being some yards off. The operator was seated in the waggon, where he had installed his apparatus; the disselboom of the waggon served as a desk for the sender to write out his despatches. So we all sent off messages, some to Cape Town, some to London, happy at finding ourselves once more in actual contact with home and with friends. At midday I finally got off, and a distance of twenty miles was accomplished before outspanning for the night. Fern Spruit was passed—of evil memory to me, as the place where three of our horses had died on the way up, and where our camp was nearly destroyed by fire. Here we picked up Major Giles, with the ox-waggon, which was to accompany us as far as Tuli. Hard by Fern Spruit is situated "Long's" Mine, from which specimens of quartz of extraordinary richness in gold have been taken. There is, however, some doubt as to whether the quartz now being worked is a legitimate reef or is not rather a " blow out." The discoverer and proprietor had dug down to a depth of only eighteen inches, and seemed to be unwilling to risk the prospect of his property by prying deeper into the earth. I expect he wanted to part with his claims for a good round sum of money to some syndicate or speculator, and take his profit at once. The specimens of quartz were sufficiently remarkable to seduce even the cautious, but with the recollection of the "Yellow Jacket" and "Golden Quarry" still fresh in my mind I passed on, not

The outspan on the Tokwe River.

even going two miles out of my way to view the mine, which had been thoroughly examined by Mr. H. Perkins. The weather for the first two days of our journey was most agreeable. The air had been cooled, the summer heat moderated by recent heavy thunderstorms, the sky was overcast with clouds, and travelling even at midday was easy for the teams and pleasant to ourselves. In appearance the bush had greatly changed since I travelled up the road. Now the vast tracts of charred and blackened ground, the result of the bush fires, were all covered with the freshest and the greenest grass. Almost all the trees were in leaf, some in flower, and the lights and shadows on the hills, on the rocky kopjes, and on the plain were of wonderful beauty and variety. We reached the Lundi on the evening of the 22nd. This stream we found greatly diminished in volume, and its passage offered no difficulty. The dangerous rocky drift of the Wanetse had been much improved by the removal of many boulders from the bed of the river, and this obstacle to travellers was traversed without mishap at sundown on the third day of the journey. From Fort Victoria to the Wanetse some effort has been made to improve the condition of the road. But a distance of eighty miles exhausted the energies of the Chartered Company's police. After the Wanetse the road relapses into a shocking condition, and stumps, rocks, deep ruts everywhere offer a profusion of danger and discomfort to the traveller. Our progress was also impeded by a marked change in the temperature. The heat became excessive; no rain had fallen south of the hills near the Wanetse river, dust enveloped the carriages in stifling clouds, and the myriads of flies almost amounted to a plague. The poor mules and horses soon showed the effect of the change, but their sufferings were added to by a total absence of young grass on which to graze during the day, and by the long distances they had often to travel in the great heat from water to water. Most of the spruits were altogether dry. One of my horses succumbed to the horse sickness. This was a horse I had purchased three months before at Victoria for a comparatively large sum, as he was reputed to be a "salted horse," and a guarantee against horse sickness was given me in his case.[1] I was now entitled to the return of the purchase-money. On the fifth day the heat became so great that we had to wait till dusk to accomplish our daily distance. No moon assisted us, and progress with "vorelopers" carrying lanterns was very slow. To an ox waggon stones and stumps and steep spruits offer little danger; it plunges along, defying all jolts and shocks. But with carriages on springs the greatest care has to be exercised, as the smashing of a wheel, or of a spring, or of a disselboom is as easy as it is irreparable. I should mention that between the Lundi and Wanetse the waggon on springs, which was drawn by oxen, and in which Major Giles was travelling, came to awful grief. Owing to careless driving in the dark across a nasty spruit the waggon was allowed to run up a high bank on one side of the passage, which toppled it over, the team drawing the forewheels and underbody of the waggon right away from the hind part. It took six or seven hours to

[1] This was most promptly repaid by the former owner.

repair the damage caused by this accident. Our night trek brought us to the Umzingwane, the ox waggon being now left far behind. A few scanty pools of brackish water in a vast bed of dry sand alone served to indicate what is at certain seasons a large and rushing river. Half-way between this river and Fort Tuli a well-filled and well-served "winkel" told us that we had re-entered regions of comparative civilization. In spite of the heat, now very great, and of the fatigued condition of our animals, we pressed on, impatient again to reach Fort Tuli, and also to accomplish a "best on record" in the way of a trek. This latter feat we did easily, and it will be long before any traveller compasses the distance between Forts Tuli and Victoria (197 miles) in a shorter or in as short a period as five days and a half, the time occupied by us.

Getting to Fort Tuli seemed like coming home again. The trek into and about Mashonaland, to which, on the 14th July, when leaving Tuli, I had looked forward with much hope and some anxiety, had been done. The truth about the country from many points of view had been fairly ascertained. Wonderful good fortune had attended us. Hardly a moment of misfortune or real trouble. Not a single moment of sickness or ill-health had been experienced by any of our large party. Day after day had glided by smoothly and pleasantly, the gipsy kind of camp life had become very fascinating, and we had had, what with sport and mining explorations, many hours full of pleasurable excitement. Nevertheless, this fact stamped itself somewhat disappointingly and sourly on my mind, that the great gold mine had not been discovered either by ourselves or by any other of the numerous exploring parties, and that the existence of any great gold mine in Mashonaland was still problematical. The Tuli river was now a vast expanse of burning sand, over which the breeze came upon you as if from the mouth of a furnace. Little threads and tiny pools of water might here and there with difficulty be detected. Fort Tuli itself and the surrounding settlement appeared in exactly the same condition as when I left it two months before. No new huts had been erected, no alterations or improvements made. No one stays at Fort Tuli who can help it; everybody passes on northwards. The strong force of Bechuanaland Border Police, under Sir F. Carrington and his officers, who made the place rather lively three months before, had retired to Macloutsie, in their own territory. Some 200 men of the Chartered Company's police are now here, but it would be difficult to determine what useful occupation they are engaged in. Major Giles brought his ox waggon into Tuli in the early morning of the 6th November. He had accomplished a still more remarkable "best on record" in the way of a trek than mine had been with mules; for with oxen he had covered the distance in exactly seven days, or, in other words, had travelled at the rate of twenty-nine miles a day. So well were the oxen looking, so little exhausted by their work, that I sold the whole span of eighteen on the day of their arrival at Tuli for 8*l.* a head. I write about these treks, for trekking is a subject of great interest in South Africa, much rivalry and emulation exists among trekkers, and rapid journeys are announced, described, disputed, canvassed, and criticized with infinite freedom and fulness. From this place I was to proceed to Macloutsie, and from thence to Palapye, where resides Khama, the redoubtable Bechuana chief. From Palapye a few days' drive would bring me through Mafeking to Vryburg, where waggons, tents, "boys," naked savages, will be all forsaken for comfortable railway carriages, civilized hotels, daily newspapers, and other similar inestimable blessings which the traveller in wild parts of the earth gets on so well without, and yet is always for a time glad to return to.

FROM TULI TO MACLOUTSIE.

Page 313.

CHAPTER XX.

LOOKING BACK.[1]

Our method of travelling—Welcome and entertainment by the Bechuanaland Border Police at Macloutsie—Palapye, the capital town of Chief Khama—Lobengula, King of the Matabele—Meditated flight of all his tribe and belongings—The Bechuanaland Exploration Company—Conversation with Khama, Paramount Chief in the Protectorate—Palla Camp—The journey to Mafeking—With Mr. Rhodes at Kimberley—The agricultural and mineral resources of the Transvaal—My advice to young Englishmen.

THE journey from Tuli to Kimberley was performed by our party rapidly from a South African point of view. Mules and horses, somewhat refreshed by a rest of two days, were inspanned an hour before daybreak on the 7th November. Macloutsie was reached at mid-day on the 9th after a pleasant drive in fine weather through an attractive country along a comparatively decent road. Our method of travelling was as follows. Aroused at about half-past three in the morning, the preparation of the coffee and the packing of the coach and "spider" occupied the best part of an hour. After trekking for two hours and a half, an outspan of an hour was necessary for the animals, and a light breakfast for ourselves was generally a welcome. Then another

---

[1] This chapter was written two months after the author's return to England. Hence its title.

two or three hours' trek brought us to our midday halt. This lasted three, four, or five hours, according to the heat of the day. If the temperature was moderate, we generally contrived to manage three afternoon treks; but often on the road south, the heat at midday was so great and the sand so heavy that only two treks, and sometimes only one, could be accomplished. The midday outspan was occupied with bathing, toilette and preparation of dinner. Our cook had been dismissed at Tuli, and the kitchen department had fallen into my hands. No very great variety in our repast was obtainable. Mutton boiled, baked, or curried, tinned soups, excellent when flavoured with Harvey or Worcester sauce, "bully beef," preserved vegetables, compose the daily meal. Fresh vegetables and fresh bread were sadly missed, but the former were not to be procured, and none of us had acquired the art of baking bread. At times these midday halts were enjoyable when we were fortunate enough to hit upon a pleasant locality on the banks of a river and under shady trees. But when, as was often the case, water and shade were conspicuous by their absence, when one was melted by the heat and persecuted by the flies, passing the hours was weary work, and the cool of the evening was anxiously longed for. Between eight and nine p.m. the day's journey terminated, thirty to forty miles having been generally accomplished. A rough supper hastily bolted, a still more hurried retirement to bed, five brief hours of slumber prepared us for another day's journey. So for the best part of three weeks we travelled, and hard travelling I found it to be. Sleep during the daytime the flies never for an instant permitted. Our two servants were worked to death; the constant unloading of the carriages for food or dressing for the night, the constant filling and refilling of the water-buckets, sometimes from half a mile to a mile having to be traversed for this purpose, made the day's toil very heavy for them. Moreover at every outspan the horses and mules were a subject of anxious care and observation. If, on being released from the harness, they at once rubbed themselves heartily in the dust or sand and commenced to graze, then all was well, but if they stood about or lay down, and were disinclined to feed, then well-grounded fears of a breakdown without a chance of assistance prevented us from being at all happy or cheerful. Before our arrival at Palla camp we had several bad half-hours on this account.

At Macloutsie we experienced the most hospitable welcome and entertainment from the officers of the Bechuanaland Border Police who have their headquarters here. The situation has been skilfully selected both as regards strategic or sanitary conditions. The camp occupies a small elevated plateau, and overlooks and commands the surrounding bush. No traveller can fail to be struck by the exceeding cleanliness and order, as well as by the excellent construction of the quarters of the officers and men. It would be difficult to speak too highly of this force. No drinking, no idleness, no slovenliness can be detected; in this lonely spot, far away from civilization, day after day, throughout the long year, the members of this force manage to occupy all their time and to keep themselves in an irreproachable condition of efficiency and smartness. No duty or errand is repugnant to the Bechuanaland Border Police. A private will start off to ride two or three hundred miles through the bush with nothing but a haversack containing biscuit, tea and coffee, and a small patrol tin. So the whole force would march, if necessary, without tents, baggage or impedimenta of any sort or kind. A wonderful *esprit de corps* animates them. Two of the officers had just returned when I arrived, from a ten days' prowl all by themselves right into Lobengula's country, entered upon partly for survey purposes, partly for the obtaining of information: a service by no means devoid of peril performed in the most light-hearted but effectual manner. Here we tarried a night and a day, lodged in comfortable huts and cheered by the comparative luxury of a well-kept mess. Major Gould Adams, the commanding officer, was in hospital, recovering, we were happy to learn, from a serious and protracted attack of fever, contracted probably when guarding the drifts of the Limpopo against the Boer trekkers. Captain Sitwell displayed the efficiency of his force in a field-day performance specially ordered for our benefit and instruction. Some two hundred and fifty men, mounted on small wiry horses in first-rate condition, scoured the bush at a gallop, to detect the presence of an enemy reported to be advancing from the north. Flying across the country in open order, they yet in obedience to bugle calls from time to time with marvellous rapidity contracted their line of advance. A small infantry detachment, dragging with it a Gatling or a Maxim gun, hurried along after the mounted men at their best speed over very rough ground. At length from an eminence the foe was descried. The men dismounting, fire repeated volleys, the Maxim and Gatling pour out a destructive torrent of projectiles, and now comes up at a gallop a 7 lbs. field-piece drawn by six horses, which quickly unlimbered looses off round after round of shell and shrapnell. The targets which represent the foe afterwards examined betray the skill and accuracy of riflemen and gunners alike. I doubted not, after witnessing this performance, that should Lobengula take it into his head to make a raid into the Protectorate, he will encounter from the Bechuanaland Border Police an uncomfortably warm reception. We were sorry to say good-bye to our hospitable hosts of the Bechuanaland Border Police at Macloutsie; but having still over four hundred miles before us, we were compelled to hurry on. Major Gould Adams most kindly lent me six fresh mules, which replaced three horses and three mules which I was obliged to leave here. Two of these horses so left, died of horse sickness almost immediately after our departure. I never heard what became of the other animals. The journey from Macloutsie to Palapye occupied four days. The country traversed was in parts most attractive, but the mid-day heat, the swarms of flies, and the heavy sand through which we had to labour were found exhausting to the teams as well as to the travellers. The last thirty miles of road into Palapye is mostly of a terrible character. The wheels of the carriages sink into the sand up to the axles, while the road is obstructed by boulders and rocks of every description and size, many of which, concealed by the sand, cannot be avoided, and the consequent jolting and straining, and peril to springs and wheels, is great. Some hours of this journeying on a very hot day took it out of us all. When Palapye was reached about five o'clock in the afternoon of the 13th November, the mules in the teams of either carriage could scarcely stir a limb. A twenty-four hours' rest was imperative. Palapye, the capital of Khama, chief of the Bangmangwato, and paramount chief in the Protectorate, stands on an elevated plateau. It is probably the most thickly populated native town in South Africa. Groups of native huts, closely packed, built without order or alignment, sheltering upwards of thirty-five thousand souls, straggle away in every direction as far as the eye can reach. A large patch of green sward, surrounded by lofty trees and covered with animals, poultry and children, reminds one strangely of an English village green. The inhabitants are all well-clothed, wear a prosperous appearance, and pay but little attention to the white traveller or passing ox-waggons. No alcoholic drink is permitted by Khama to find its way into his territories, or under any circumstances to be sold in his towns. The penalties for violating this law are most severe, and are severely enforced. Constant raids by Khama's police, sometimes led by Khama in person, swoop down upon all prostitutes and immoral persons, who are forthwith banished from the town. Khama governs justly and severely, but without cruelty. Human life is, I believe, never taken. His authority is purely despotic, undisputed, unrestrained, but exercised with wisdom has secured for him the affectionate respect of his people. He is the most powerful chief in South Africa with the exception of Lobengula, King of the Matabele, nor would it be possible to predict with any assurance the result of a conflict between these two potentates. The soldiers of the Matabele army are possibly more brave and ferocious than the Bangmangwato, but the latter possess a considerable advantage in their numbers of mounted warriors, of which the Matabele are entirely destitute. In the event of Lobengula attacking the British settlers in Mashonaland, Khama could almost certainly be persuaded to go at him and to effect a powerful diversion. Khama and the British Government have reciprocally benefited each other. The Protectorate was submitted to and English authority acknowledged without resistance, mainly owing to the friendly attitude of Khama. On the other hand, his authority has been strengthened by British assistance and good offices, and, confident in British support, he no longer fears his enemy Lobengula. As to the probable conduct of Lobengula in the future, I procured some interesting information from an English gentleman long resident at Palapye, whose father dwells at Baluwyo. In his opinion Lobengula has long meditated a flitting with all his tribe and belongings, into the country north of the Zambesi, where he calculates to carry on with ease an unrestrained and exterminating war of conquest. The great difficulty in the way of this policy is the transportation of the immense herds of cattle and sheep, the property of the monarch and his people, across the broad and rapid Zambesi. But my informant thought that, as white settlers and merchants multiplied, and as British influence and domination increased, Lobengula would get more and more uneasy, more bent and resolute on his policy of migration towards the north. But my informant felt certain that before any such migration actually took place, Lobengula would make himself, or would allow his young men to make, a last dying effort as it were against the white people, when much bloodshed and massacre might occur. It is not, however, likely that the British authorities will not obtain ample notice beforehand of the imminence of any such attack. Khama possesses in his rival's city many and various channels of sure information. Nor are the Administrator of Bechuanaland or the police force by any means poorly provided for in this respect. But I expect that for a long time yet it will be necessary for the British settlers in Mashonaland and north of the Crocodile river to exercise the utmost caution, not only as to their conduct towards the Matabele, but also as to the preparation of measures for concentrated resistance in the event of an outbreak of savage fury.

The Bechuanaland Exploration Company, which does a large and profitable trading business throughout these parts of Africa, has its northern headquarters at Palapye. From their agents we experienced the utmost kindness; nor among the least of the luxuries they offered was a brandy and soda, which, besides being the first I had been able to get for many weeks, was, to a traveller suffocated by heat and choked by dust, sweeter than any heavenly nectar. I must add that the brandy bottle was produced from a recess under the bed, in one of the huts occupied by the agent, where, in deference to Khama's teetotal proclivities, it was carefully concealed. Nor can I omit to mention another great luxury here enjoyed for the first time for more than five months, to wit, a night's repose between a pair of sheets. The Bechuanaland Exploration Company have here a large and well-supplied store, where we procured many articles of which we stood in need. In the morning I witnessed a curious spectacle. Many hundreds of Khama's people who had been employed by the South African Chartered Company during a period of four months in laying the telegraph wire through Mashonaland were now paid off. A large quantity of gold and silver coin had been brought up by an officer of the Bechuanaland Border Police from Mafeking, and was by him distributed to this immense crowd with the utmost order, accuracy, and general content. Some of the sub-chiefs received very considerable sums of money from their people, ranging as high as 150*l*. The store did a roaring trade, and till evening the natives kept passing our encampment on their way home, laden with blankets and beads, and very many I noticed carrying brand-new Martini-Henry rifles. Palapye is a great emporium for horns, skins, karrosses, and native curios, and I added some fine specimens of these former articles to the collection I had already formed in Mashonaland. In the evening of the 14th November, about half an hour prior to our departure, my servant came to inform me that the chief Khama had come to visit our encampment. I hurried to welcome him, and found myself in the presence of a tall, slight man of apparently about forty years of age. Khama is, I believe, a good deal older. He was dressed in a suit of woollen stuff of English make, and looked like a coloured manager of a factory in India, or of a cotton plantation. A very intelligent countenance, an agreeable and kind expression, an erect attitude and dignified manners mark the monarch, the minister and the father of the people. Our conversation on commonplace topics, lasting about a quarter of an hour, was carried on by the interpretation of Mr. Secker, agent of the Bechuanaland Exploration Company. At the close Khama graciously intimated that he would like to make me a present, and inquired if I would accept one. I replied that any memorial of him would be most welcome and valuable to me; he then took his leave, galloping off on a fine bay horse which he rode with grace, followed by his equerry, and looking, I thought, in that position a king all over. Shortly after a messenger arrived, bringing me from the chief a large karross, made of leopard skins of a quality and fineness such as a great chief would alone possess or be able to procure. I sent Khama in return a large silver flask, which I told him was my "water bottle," and which I hoped he might sometimes use in his hunting expeditions, as a souvenir of an English traveller and friend. Then we departed for Palla Camp. The road leaving Palapye for the south is even worse on account of deep sand and rocks than the approach before mentioned. To save our mules during a trek of twenty miles through this ground, we had inspanned into the "spider" and coach two teams of oxen. All through the night we travelled, our mules driven along slowly behind at their ease. At daybreak, after an outspan, we resorted to our mules, now much rested and refreshed, and made good progress through some beautiful bush country, until at noon on the 17th November we reached Palla Camp. Here is a telegraph station, a small police detachment and a good store. Leaving Palla early the next morning, about 10 a.m. we met the up-country mail, in which was Mr. Harber, the mail superintendent of the Bechuanaland Exploration Company, who had been very kindly sent by the agents of the company, to take charge of our party from here, and to supply us with fresh relays of mules. At first when reaching the post stations, Mr. Harber could only provide an occasional fresh mule, and we toiled along rather wearily and very slowly, owing to the heavy sand, to Machudi's Kraal or Lenchwe, as it is sometimes called. On the day before reaching this place, Mr. Harber and I saw from the "spider" ahead of us a large snake lying in the road. Out we jumped, I with my revolver he with a formidable "sjambok" to slay the reptile. This snake made off into the bush with such celerity and such twistings that I discharged all the barrels of my revolver at it in vain. Mr. Harber, however, arrested its progress with a blow from the thong of the sjambok. It reared up, opening wide its mouth at us and hissing, a grand object. Fortunately, owing to the length of the sjambok, it could not reach Mr. Harber, who almost immediately laid it low with a well-directed blow, the thong catching it tight round the neck. A bite from this most venomous of African snakes would have been certainly fatal in less than an hour. It was a puff-adder of immense size, measuring seven feet, with a body thicker than my wrist. At Machudi's a native kraal, inhabited by a tribe only second to Khama's in influence and numbers, and governed by a chief of evil reputation, our travel troubles terminated. Here our heavy coach was to be left. Major Giles and the doctor would travel in a smaller and lighter vehicle, which, together with the "spider," was to be taken along from here entirely by the company's teams. We reached Machudi's not a moment too soon; our own animals could not have gone another yard without a rest of several days. But it seemed as if fate was against us. The river Notwani here to be traversed was found high in flood and perfectly impassable, in which state it might remain for a week or more. I went to bed with a heavy heart, anxious about the future of our journey south. In the morning I was cheered by the news that the river had rapidly subsided, and that Mr. Rhodes and his party, also being conveyed by the Bechuanaland Exploration Company, had passed us, crossed the river, and gone on in the night. From here we proceeded rapidly and gaily, finding fresh teams of mules every two or three hours, passing through Gaberones, a station of the Bechuanaland Border Police, where we received the usual most bountiful hospitality, through Ramoutsa, an important post station, across beautiful green pastures, through succeeding forest, wood and bush, everything looking bright and verdant and glistening, owing to recent heavy rains, through Ramatlabama on to our goal, Mafeking, which we reached in the afternoon of Monday, the 23rd November.

At Mafeking, Dixon's Hotel, an establishment of the greatest merit, soon enabled us to forget the troubles and fatigues of our long journey of a thousand miles from Fort Salisbury. The railway will soon be extended to this pleasant and attractive settlement. It lies in the centre of a boundless grassy plain, at this time of year (November) very green and fertile. Bechuanaland is destined, I imagine, in time to become the great ranching ground of South Africa. Skilful engineering in the placing and making of dams, would by storage overcome the only obstacle, viz. the scarcity of water. We had to remain at Mafeking two days. A drive of one hundred miles still separated us from Vryburg, the railway terminus. The passage of Mr. Rhodes and his party had absorbed the coaching resources of the Bechuanaland Exploration Company. The mail service up country as far as Tuli is excellently carried on by this company. Passengers are also conveyed by them, the distance from Vryburg to Tuli, upwards of five hundred miles, being compassed by the mail coach in less than eight days. The company have large numbers of mules, all kept in first-rate condition, and have constructed post stations at intervals of twenty miles along the road, where are wells and stores of forage. If I were going again to Fort Salisbury, I should make arrangements with this company for my journey, instead of resorting to the large, very costly and cumbrous expedition which ignorance of the country let me in for. A traveller, by making use of the present mail service and by procuring from the company a special service north of Fort Tuli, could journey from London to Fort Salisbury and back in a period of four months. Of course if the Beira Pungwe railway were constructed the journey could be accomplished in half that time. The distance between Mafeking and Vryburg was covered in a day. Starting at 3.30 p.m., the relays and mail teams taking us along with unusual rapidity, we reached Vryburg at 8.30 in the evening. Oh! the comfort and luxury of the railway, after seven months of travelling in coaches and waggons. A week was pleasantly passed at Kimberley, where I was the guest of Mr. Rhodes. No change could be noted here. The concentration of the diamond industry into the hands of a single company has cramped the development of this town. But there is there a hospitable and amiable society, and the most comfortable and well-managed club I have ever come across in my numerous travels. Captain Tyson, the secretary of this club, is a perfect providence to the English visitor. So once more in Capetown, where I whiled away three weeks waiting for Mr. Perkins, the mining expert, to rejoin me from Johannesburg where he had been engaged in a second minute examination of the gold-field of Witwatersrand. The rest and the comparative idleness after so many weeks of hard and rough travel, and above all the gracious hospitality extended to me by his Excellency the High Commissioner and Governor, were enjoyable beyond description. Time was now ample for reflection and retrospect, nor were materials for such wanting. The following problem continually presented itself to me: How could the paucity of British population in the Cape Colony, and in South Africa generally, be accounted for?

Soil and climate equal to that of Australia, vastly superior to that of Canada, should have attracted a constant stream of emigrants, either to the Cape, Bechuanaland, to the Transvaal or to Natal. Such for some reason or other has not been the case. Possibly the Dutch element so predominant throughout South Africa is unfavourable to rapid enterprise, possibly the large amount of cheap native labour conflicts with the attainment of a very high standard of colonial prosperity and strength. Whether it be so or not, the question presents itself for study. In Australia and Canada many millions of population, mainly drawn from British sources; in South Africa from the Cape to the Zambesi, a territory of vast expanse, with miles of fertile pastures most suitable to cattle and sheep, with acres of land capable of producing abundant crops of grain, with forests giving most valuable and excellent timber, with mines of every metal, and with large deposits of coal, is inhabited at present by about half a million of white people, not more than two-thirds of which are of British origin.

The most sanguine dreamer can hardly over-estimate the agricultural and mineral resources of the Transvaal. Before the end of the year the railway will have superseded the ox waggon, Johannesburg and Pretoria will be connected with the railway systems of Cape Colony and of Natal. This should produce a rapid and large increase of population and of mining industry. Probably in the history of mining, no gold-field more important than the Witwatersrand has ever been discovered. When I passed through Johannesburg in June, 1891, the monthly output of gold from its mines was 54,000 ounces. At the time of writing this has risen to 86,000 ounces. Three causes will contribute to sustain and swell this remarkable development. **1. The** general introduction into the mines of the compressed air rock-drilling machinery, and a consequent large increase in the amount of auriferous ore extracted, together with a saving in the charge for labour. 2. The marked success of the chemical processes **for treating** tailings, with a consequent large **increase in** the amount of gold actually won. 3. The construction of the railway to Johannesburg, with a consequent **large** decrease in working expenses, and in the **cost of** living. There is now before **many** if not all of **the** Witwatersrand mines **an** amount of auriferous **ore** practically **in** sight which can exhaust the energies of **at** least another generation of men. **Of** the silver deposits near Johannesburg no absolutely definite and **precise** allegation can be **made.** Their promise is good, and almost warrants **the** speculation **that** some day **the silver** mining industry will rival **if not** surpass in importance the **gold-mining** industry **of the Randt.** It is to the **Transvaal** wealth that **I look** for the attraction which **may ere long thickly** populate South Africa. It is impossible **not to** regret that a policy **as** some say of prudence, **as others say** of cowardice, compelled Great Britain **to give** up her direct **authority** over this land, but the **riches of the** world are there in abundance, nor is it in the power of a feeble, corrupt and almost insolvent Boer Government to prevent or to delay for long these riches being largely distributed among mankind. Pages I could write in praise of South Africa, but fortunately want of space arrests me. To the young, vigorous and versatile British emigrant, I can recommend the country as a place where the means of ease and affluence can be acquired rapidly, if only fortune smiles; to the traveller in search of health, distraction, amusement, sport, beauty of scenery, excellence of climate, I can recommend it as being the region of the world most favoured by nature, either for the residence or the industry, or the wanderings of man.

THE END.

INDEX.

ADDERLEY Street. Cape Town, 19.
Africa, coaching in, 53-4; equipment of an expedition for, 117-121; hunting the lion in, 161-2; cost of a six months' hunting expedition in, 216-8; a genuine stickfast in, 221-2, 257.
Agriculture in Mashonaland, 276-7; at Hartley Hill, 284; in Transvaal, 328-9.
Alcohol in the Transvaal, 92; at Palapye, 319.
"Alice" reef, Mazoe Valley, 205-6.
Amatongaland, 28.
Amusements on the *Grantully Castle*, 11, 12.
Antelope, 150-5, 158, 165-170, 183, 199, 211, 214, 218, 220-6, 228, 233, 249-262, 267; see also blesbok, buck, harte-beest, gazelle, koodoos, springbok, &c.
Armament required for hunting expedition in South Africa, 217.
Athletic sports on the *Grantully Castle*, 12, 13.
Auction at Fort Salisbury, 247, 284-6.

"BABOON," the (Lee's boy), 147, 150, 153-9, 225-9, 232, 253-9, 262.
Baboons, 243, 258.
Bads loop, 98.
Baluwyo, 320.
Bangmangwato tribe, **318, 319**.
Basutoland, government **of, 27**.
Beale's Camp, 221, 226.

Bechuanaland, government of, 27, 326; native reserve in, 50-52; Border Police of, 105-8, 124, 129, 132, 136, 142, 144, 191, 300, 311, 315-7, 322; Sir Charles Warren and, 124.
Bechuanaland Exploration Company, 321-6.
Beit, Mr. Alfred, 143, **191, 210**, 233, 242-3, 265, 270, 272.
Benett-Stanford, Mr., **7-9**.
Benkes, Mr., 86-7.
"Birthday" mine, the, 99.
Blandy and Co., **Madeira, 9**.
Blesbok, 75, 76.
"Bless," horse named, **184, 186**, 189.
Blue ground extracted at Kimberley, 41.
Boers in the Transvaal, 22-25; at Johannesburg, 60-64, 72; and game, **75**; the Transvaal Parliament, 81-88; idea of justice, 88-92; **trek** by, into Mashonaland, **88-9**, 93, 108-110; and Swaziland, 92-4; **as** farmers, 94-5; and wells, 102.
Borrow, Mr., 218, 239, 243, 296, 298; see also *Johnson, Heaney and Borrow*.
Botanical Gardens at Lisbon, 7; at Cape Town, 18.
Botany, garden **at** Madeira, 8-9; the petuna, 258.
Bread a luxury in the bush, 180.
Breakwater at Cape Town, 19.
British Chartered South African Company, 22, 26, 104, 105, 108;

police of, 111-112, 124; at Fort Victoria, 190-192; 196-7, 205, 231; huts of, 206; at Fort Salisbury, 282-9; police of, 299, 300.
Bubjane River, 156, 175, 176.
Babye River, 150.
Bucks. 142, 144, 145, 173, 211, 214, 217, 218, 220, 228, 250, 266, 294.
Bufflesdorn Mine, 57.
Buildings at Fort Salisbury, 282.
Bullock-vehicle, Madeira, 9.
Bultfontein Mine, see *De Beers Co.*
Buluroyo, 268.
Bush fires, 187-8, 308.
Byl, Mr. Van der, 302.

CAMERON. General, 28-31.
Camp by moonlight, our, 144.
Camp fire concert at Fort Tuli, 124-5.
Cape de Verde, 9.
Cape Town, arrival at, 15-17; buildings in, 17-19; quietness of, 19, 20; environs of, 21; as a coaling station, 28; defences of, 29-30; garrison at, 31; departure from, 32-33.
Capper, Captain, and the new magazine rifle, 113-5.
Carbolic oil, usefulness of, 102.
Carrington, Sir Frederick, 106-8, 110-112, 125, 142, 144, 252, 311.
Cattle disease, 51.
Cattle of Boers, 94; at Fort Victoria, 190.
"Charlie," shooting pony, 158, 186.
Chimpanzees, 7.
Chlorination process, the, 66-70.
Churchill, Lord R., journey to Cape Town, 1-16; Cape Colony, 17-33; at Kimberley, 34-49; at Johannesburg, 50-78; and the Transvaal, 79-95; the journey to Fort Tuli, 96-125; and the lions, 158-174; the journey to Fort Victoria, 175-192; from Fort Victoria to Fort Salisbury, 193-211; sport in Mashonaland, 212-233, 246-262; the Mazoe Valley gold district, 234-245; and wealth of Mashonaland, 263-275; at Fort Salisbury, 276-294; the journey home, 295-327; advice to emigrants, 328-330.
Climate of Cape Town, 17, 21; of the Karroo plain, 35-6; of Transvaal, 68, 72, 81, 124; of Pretoria, 82; at Fort Tuli, 124; of Mashonaland, 198-9; at Fort Salisbury, 206; of Mashonaland, 293, 301.
Coaches and coaching in South Africa, 52-5, 298-9.
Coal mines near Johannesburg, 74.
Coaling station, Cape Town as a, 28, 29.
Colquhoun, Mr., 200.
Concerts at Fort Tuli, 124-5; at Palla Camp, 131.
Concessions in the Transvaal, 64.
Cooking of venison, the, 220-1.
Corruption in the Transvaal, 64.
Cost of a six months' hunting expedition in South Africa, 216-8.
Coventry, Honourable Charles, 119, 153, 233, 265.
Crocodile River, see *Limpopo River.*
Crocodiles, 130.
Cruelty of Boers, 88-92.
Cyanide of potassium process, 66, 69, 70.

DAMARALAND, 28, 52.
Dartmouth, 1, 5.
De Beer, Mr., 86-7.
De Beers' Company, the, at Kimberley, 38-48, 191.
Deer, preservation of, 76.
Defences of Cape Town, the, 28-31.
Desolation, a time of, 138-9.
Diamond industry at Kimberley, 36-49.
Docks at Cape Town, 19.

INDEX. 333

Doctors, lack of, in Mashonaland, 202-3.
Dogs, advice about, for South Africa, 102; our, 160, 168.
Donkeys, 266.
Dutch in Cape Town, the English and, 22-25, 328.
Dutch Parliament, the, at Pretoria, 83-88.
Du Toits Pan mine, see *De Beers Co.*
Dynamite, a monopoly, 64.

EDGELL, MR., 116, 117, 134, 136, 143, 230, 265.
"Eiffel" district, the, 265, 270.
Elands, 199, 218, 225, 233, 239, 259-262.
Elebi, 135.
Electric light in the De Beers Mine, 48; in the Robinson Gold Mine, 65.
Elephants, 147.
Emigrant, Mashonaland for the, 237-8; South Africa for the, 330.
English and Dutch in Cape Town, the, 22-25.
Exploration Company Syndicate, 244-5.
Expedition, the, composition and equipment, 116-121; sale of effects of, 284-6.
Eytings, 99.

FAIRYLAND, a veritable, 103.
Farmer, the Boer as a, 94-5, 192.
Feathered game in the Transvaal, 76.
Fern Spruit, 185, 186-8, 306.
Ferreira, Col., 109, 191.
Ferreira Mine, the, 69.
Fever at Palla Camp, 129; at Lundi River Camp, 181-3; at Fort Victoria, 190; in Mashonaland, 203, 237-8.
Fire on board the *Grantully Castle*, 13-15.
Fires, Veldt, 187-8, 230, 308.

"Fly" (grey gelding), loss of, 182.
Flying fish, 10.
Footpads in Johannesburg, 60.
Fort Charter, 191, 195-8, 200-203.
Fort Salisbury, 193, 200-8, 211, 218, 230-6, 238, 246, 258, 274, 281-298.
Fort Tuli, 96, 109, 110-116, 142, 291, 310-13.
Fort Victoria, 189, 193, 201-3, 280, 291, 300-310.
Fort Wynyard, 30.
Fraser, Messrs., rifle made by, 118.
Frere, Sir Bartle, 24.
Funchal Bay, 7.

GABERONES Station, 325.
Game in Transvaal, 102; on South Africa veldt, 150-2.
Garrison at Cape Town, the, 31; at Fort Charter, 202.
Gascoigne, Major, 103.
Gazelles, 225.
Gideon (boy), 149.
Giffard, Mr., 289.
Giles, Major George, 3, 4, 116-7, 122, 130-132, 136, 145, 153, 170, 174, 176, 181, 186, 193, 265, 266, 298, 306, 309, 311, 324; accident to, 230; and the horse sickness, 121-3.
Giraffe, 133, 173, 218.
Gladstone, Mr., 5, and the Transvaal War, 23-25.
Gold near Hartley Hill, 200; in Mashonaland, 207-211, 236, 271, 277-281; see also *Mazoe*, &c.; round Fort Victoria, 302-3.
Gold-field of Witwatersrand, 327-9.
Gold mines in Johannesburg, 59, 63-73, 79-81.
"Golden Quarry" mine, 243, 297, 307.
Goold-Adams, Major, 108-10, 316, 317.
Government House, Cape Town, 18.

Government buildings at Pretoria, 83.
Governments in South Africa, various forms of, 25-28.
Graham, Mr., 264.
Grahamstown Mine, the, 71.
Grantully Castle, voyage in the, 5-16.
Guns, breech-loading at Cape Town, 29-31.
Gweebi River, 239.

HAMPDEN, Mount, 207, 211, 239, 241, 295.
Harber, Mr., 323, 324.
Harris, Dr. Rutherford, 288-290.
Hartebeests, 140, 155, 199, 214, 219, 222-6, 232, 233, 250-4, 266, 268.
Hartley Hill, gold district of, 200, 208, 209, 236, 237, 246-7, 253, 258, 262, 263-274, 279.
Hex River, 34; Pass, 34-5.
Hippopotami, 258, 294.
Honey bird, the, 147.
Hopley, Mr., 284.
Horse-racing at Fort Salisbury, 289, 290.
Horse sickness in Africa, 51, 121, 127-8, 136, 176-7, 181, 183, 186, 190-2, 309, 315.
Hotel accommodation, in the Transvaal, 55-6; at the Warm Baths, Pretoria, 98, at Pietersburg, 101.
Hot springs near Worcester, 34; near Pretoria, 98.
House-breakers in Johannesburg, 60.
Hunting in South Africa, 212-8.
Hunyani River, 204, 211, 212, 218-19, 233, 238, 265.
Huts of the B.S.A.C.C., 206; of Kaffirs, 255.
Hyænas, 133, 160, 218.

IPAGI River, 145.
Iddesleigh, Lord, 20.
Illicit diamond buying in South Africa, 45-7.

Insects. Ants, 258; Black flies, plague of, 263; tsetze fly, the, 160, 213, 265, 266, **298**; caterpillars, 225.
Inspanning, the **business of,** 143.
Invalids, South Africa and, 15, 16.
Irish Land Question Bill, 2.

JACKALS, 160, **211, 218, 229.**
Jahshaan, 102.
Jamieson, Dr., 206, 218.
Jantje, a native, 89-91.
Johannesburg, 49; the journey to, 53-7; description of, 58-60; taxation in, 61-2; government at, 63-4; mines at, 65-75, 79-83, 328-9.
Johnson, Heaney and Barrow, Messrs., 218, 239, 242, 264, 272, 282-3, 298.
Joubert, General, 84, 88, 109, 110.
"Jumbo" mine, the, 243.
"Jumpers" Gold **Mine,** 70.
Justice, the Boer's idea of, 88-92.

KAFFIR, maltreatment of a, 88-91.
Kaffirs, 103; women, 128-9; waggons, 137; kraals, 255.
Karroo, plain of the, 35-6.
Kenilworth, model village, at Kimberley, 47.
Khama, Chief, 145, 312, 318-323.
Kimberley, 119, 313, 327-8; diamond industry at, 36-49.
Kimberley Mine, see *De Beers Co.*
Klerksdorp, 56-8.
Knollys, Colonel, 28.
Koertze, Mr., 76.
Koodoos, 133, 145-8, 150, 155, 160, 164, 172, 176, 218, 225, 232, 259; see also *antelopes.*
"Kopjes," rocky, 179, 206, 249.
Kraals of natives in Mashonaland, 204, 241, 254.
Kruger, President, 84-88, 93, 110.

LANGE, Mr. A. E. de, cruelty of, 88-91.
Langlaate Estate, the, 68.

INDEX.

Lanyon, Sir Owen, 24.
Laurie, Captain, 96, 105, 107.
Lee, Mr. Hans. the hunter, 120, 144-8, 150-160, 176; 182-5, 211, 212, 217, 233, 248-251, 254-262, 267; and the lions, 161-174.
Leonard, Captain, 110.
Leopards, 36, 213, 275.
Library, Public, at Cape Town, 18, 19.
Licenses in Fort Salisbury, 292.
Limpopo River, 103, 106-108, 129, 131, 132, 135, 142, 212, 316.
Lion Camp, 158-174.
Lions, 132, 155, 157, 160-172, 199, 204, 213, 217, 264, 267, 302.
Lipokwe River, 137.
Liquor traffic at Fort Salisbury, 292.
Lisbon, 6-7.
Livestock in the Transvaal, 51.
Lobengula, Chief, 125, 145, 204, 268, 316-320, see also *Matabele*.
Loch, Sir Henry, 2.
Logan, Mr. J. D., 35-6.
Lo-Magundi district, 271, 279, 281.
Long's Mine, 306.
Lost in the veldt, 138-141, 252-3.
Lotsani River, 134, 136.
Lottery on board *Grantully Castle*, 11.
Lundi River, 174, 176, 179, 181, 184, 302, 308-9.

McArthur-Forrest process, the, 66-70.
Machudi's Kraal, 324-5.
Mackay, Mr., 116, 117, 122, 131, 134, 136, 143, 153, 265, 266, 270, 280.
Macloutsie, 136, 311-318.
Madeira, 7-9.
Mafeking, 49, 126, 312, 325-6.
Magazine rifle, the new, 112-115.
Magistrates in Mashonaland, 290.
Mahalopsie River, 131.
Majuba Hill, 23, 24.
Makala tribe, 153.
Malarial fever, see *Fever*.

Mammoth River, 271.
Manicaland, 104; gold district, 207, 236-7, 269, 274, 280, 294.
Mariko River, 128.
Marico district, the, 192.
Maripi, 128.
Marks & Co., estate of, 76-8.
Marriage in Mashonaland, 248-9.
Martini-Henry rifle, the, compared to the new magazine rifle, 113-5.
Mary Pioneer mine, the, 297.
Mashonaland, 2, 4; and the Boers, 85-6, 93; wealth of, 175; the best part of, 183; climate and soil of, 198-9; from a mining point of view, 209, 293; emigration in, 237-8; wealth and fertility of, 269, 271, 276-281; natives' dress, 286; postal communication in, 290.
Massi Kessi, skirmish near, 104-5, 300.
Matabele, raids of the, 202, 204, 205, 241, 321; see also *Lobengula*.
Matabeleland, 25, 26, 28, 52, 86, 110; women, 128-9; tree in, 160.
Matchless mine, the, 279.
Matjesfontein, 35.
Matlaputta River, 136.
Maunde, Mr., 179, 180.
Maxim gun at Fort Tuli, 109, 112; at Macloutsie, 317.
Mazoe River gold district, 200, 208, 210, 233, 234, 237-246, 269, 274, 279, 295-7.
Menu at hotel in Transvaal, 55-6.
Mineral resources, of Matabeleland, 25; of Transvaal, 100, 328-9; of Mashonaland, 278-281.
Mines, see *De Beers Co.*, *Robinson Co.*, *Kimberley*, &c.
Mines near Kimberley, 57; in Johannesburg, 58-60, 65-75.
Mining in the Zoutspanburg district, 99, 100.
Mockell, Mr., 117.

Monkeys, 7; see also *baboons, chimpanzees*.
Montgomery, Sergeant - Major, 289.
Morier, Mr. Victor, 103-5.
Morrison's store, 105.
Mountains, some miniature, 179-80.
Mount Marias mine, 99.
Mules, driving a team of, 53.
Mules, our, 143, 156, 176-7; habits of, 148-9, 182, 191-2, 199, 203, 308, 315, 317.
Murchison district, 99.
Myberg, Mr., 117, 149, 157, 166-9, 179, 182, 184.

NATAL, Government of, 26.
Native market, 153-4.
Native reserve in Bechuanaland, 51.
Natives as servants, 194-5, 247-8.
Natural History Museum, Cape Town, 18, 19.
Nelmapius, Mr., 100.
Notorious diamond thief, a, 46-7.
Notwani River, 325.
Nylstrom, 98.

ORANGE Free State, 27.
Ornithology—
    Bittern. 77.
    Bustards, 36, 76, 225.
    Cranes, 76, 78.
    Doves, 101.
    Duck, 78, 127, 130.
    Eagle, 77.
    Guinea fowls, 101-2, 137, 139.
    Honey-bird, the, 147.
    Koran. 36, 55, 77, 78.
    Partridges, 36, 55, 77, 78, 101.
    Pheasants, 101, 137, 139, 142, 144.
    Pigeons, 77.
    Plovers, 55, 77-8.
    Quails, 36, 77-8.
    Snipe, 77-8.
    Teal, 130.
    Vultures, 55, 75, 229.

Ornithology (continued)—
    Wild fowl, 77.
    Wild turkey, 78, 239.
Ostriches, 183-4, 199, 218, 239, 259, 260.
Outfit necessary for hunting expedition, 216-7.
Oxen, loss of, 63; our, at Tuli, 123, 143, 146, 184, 230; habits of, 148-9.

PAARL, old town of French origin, 33-4.
Paddington Station, 1, 5.
Paddington man, a, 35.
Palapye, 312, 318-323.
Palla Camp, 129, 315, 323.
Palmitsfontein mine, the, 99.
Panouse, Count de la, 243.
Papenfu, Mr., 284.
Paris Exhibition, diamond exhibited at, 42.
Paritj, estate near, 76.
Pelapswe, 136.
Pennefather, Colonel, 104, 207.
Perkins, Mr. H. C., 4, 97, 210, 211, 234, 235, 238-247, 253, 265-7, 271, 274-5, 287, 298, 307, 327.
Personnel of Expedition, 116-121.
Pietersburg, 99-101.
Plains in Africa, see *Veldt*.
Police, at Johannesburg, 60; at Fort Salisbury, 290; the Bechuanaland Border, 105-8, 124, 129, 136, 142, 144, 300; of the B.S.A. Co., 299, 300.
Politics in England, 2; in South Africa, 25-28.
Poll tax in the Transvaal, 62.
Pondoland, 27.
Port Elizabeth, 19.
Portuguese, skirmish with, near Massi Kessi, 104-5, 110, 300.
Postal Communication in Mashonaland, 290.
Potchefstrom, 57.
Power of President Kruger, 87-8.
Preservation of deer, 76.
Pretoria, 82, 97, 328; taxation in,

61-2; Dutch Parliament at, 83-88.
Prospecting Mashonaland, 278.
Providence Gorge, 189.
Provisions required for a six months' hunting expedition, 217-8.
Pullen's "Winkel" in the Transvaal, 55-6.
"Pulsator" machine, De Beers Mine, 42.
Pungwe River and route, 104-5, 209, 279, 294, 298.

QUAGGAS, 147, 150, 165-171, 214, 218.

RAILWAYS in the Transvaal, 49, 50, 63, 73, 80, 328-9.
Rains, heavy, 145, 177, 237-8, 300; scarcity of, 136.
Ramatlabana, 126, 325.
Ramoutsa, 325.
Randt gold-field, the, 71, 72, 74.
Rayner, Surgeon Hugh, 4, 126, 144, 186, 203, 265-8, 298, 324.
Reptiles, see *Snakes*.
Reserve for Natives in Bechuanaland, 51.
Rhinoceros, 294.
Rhodes, Mr. Cecil, 2, 22-4, 95-7; 200, 288, 293-7, 305, 325-7; and the De Beers Co., 38.
Rhodes's Drift, 105, 107.
Rifle, the new magazine, 112-5.
Roads, in the Transvaal, 54, 63-4; between Fort Victoria and Fort Charter, 299, 300.
Robinson Gold Mine, the, 65-9.
Rolker, Mr., 209, 234-5, 242, 246, 247, 253, 265, 287.
Romilly, Mr. Hugh, 191.
"Ruby," horse named, 186.
Rustemburg Goal, 88-91.
Rylands and Fry, Messrs., 173, 176.

SALARY of members of Dutch Parliament, 87.
Salisbury, Lord, 29.
Salisbury Gold Mine, the, 70.
Sandpits, 127.
Sanitary Board at Johannesburg, 61-2.
Sapte, Major, 103-4.
Saroe River, 258.
Saur, Dr., 132-134.
Search room at Kimberley diamond mines, 45.
Seeker, Mr., 322.
Selous, Mr., the hunter, 18, 207.
Semalali River, 137.
Sequana, 128.
Servants, our native, 194-5, 247-8.
Shave by a Hindoo barber, 134.
Sheep in the Transvaal, 51; sheep scab, 51.
Shepstone, Sir T., 24.
Shippard, Sir Sydney, 52.
Shooting in South Africa, 212-8.
Silika, 132.
"Simmer and Jack" gold mine, 70.
Simon, Dr., 69, 70.
Simon's Bay, defence of, 30.
Sinclair, Mr., 127.
Sitwell, Captain, 316.
"Skoff," 132, 134.
Slater, Mr., 284, 289.
Smitsdorp, 99.
Snakes, 213; a cobra, 227; scorpions, 275; a puff adder, 324.
Soil, in the Transvaal, 80-81; of Pretoria, 83; of Mashonaland, 198-9, 238, 276-7; at Fort Salisbury, 206; at Hartley Hill, 263.
South Africa for invalids, 15, 16; trekking in, 312; for emigrant, 330.
"*South Africa*," a number of, 128-9.
South African Republic, 22, 24.
Speculation in gold mines, 281.
"Spider," the, travelling by, 100, 105, 176-8, 182.
Sport on the Karroo plain, 36; in the Transvaal, 75-8; with Sir F. Carrington, 142-4; on the veldt, 147-152; see also *Zoology*, &c.

Z

Springbok, 36, **75**, 76, 78.
Stauhopo, Mr. (Secretary of State for War), 30, 32; and the new magazine **rifle**, 112-115.
Stickfast, **a genuine** African, 221-2, 257.
Stock Exchange, London, and Johannesburg gold mines, 59.
Suchi River, 136.
Sugar Loaf Mountain, 179, 180.
"Susanna" reef, Mazoe valley, 295-6.
Swallows, 10.
Swaziland, government of, 27; the Boers and, 92-4, 110.

TABLE BAY, 15; defences of, 29.
Table Mountain and Bay, 15, 17.
Tagus River, 6, 7.
Tatagora River, 239.
Taxation in the Transvaal, 61-2.
Tati gold-fields, 49, 50.
Taxes in Fort Salisbury, 292.
Telegraph wire at Macloutsie, 136; to Fort Victoria, 145, 305-6; at Fort Salisbury, 291.
Telephones fixed in the De Beers Mine, 48.
Temperature at Fort Tuli, 124; at night, 126; at Hartley Hill, 264, 268.
Terra Santa, Island of, 7.
Theft of diamonds at Kimberley, 44-7.
Thief, a notorious diamond, **46-7**.
Thorns, 130.
Ticks, dogs and, **102**.
Tiriki (servant), **248-9, 303-5**.
Tokwe River, 302.
Towlu Mount, **155**, 157.
Transvaal, **the**, 27, 50, 52, 81-2; hotel accommodation in, 55-6; Silver Mi**nes** Co., 74; deer and feathered game in, 76-8; government at Pretoria, 83-88; Boer justice, 88-91; natives in, 92; Boer farmers in, 94-5; mineral resources of, 100, 328-9.
Transvaal War, the, 22-25.

Trees, at Madeira, 8, 9, 130; in Transvaal, 73, 81, 97; a, in Matabeleland, 160; in Mashonaland, 240-1; "Cream of Tartar" tree, 103; elephant fruit tree, 147; mahogany tree, 155; snake tree, 159; mogundi tree, 224; makoona tree, 254; wild fig-tree, 258.
Trek by Boers into Mashonaland, 85-6, 93, 108-110; from Vryburg to Tuli, 121-2; through the bush, 146; a record, 310, 311, 314.
Tuli River, 96, 143; see also *Fort Tuli*.
Turner, Captain, 190.
Tye, Major, 110.
Tyson, Captain, 327.

UMFULI RIVER, 200, 208, 221, 225, 263, 271.
Umfuli River, the upper, 203, 204, 215, 238.
Umjinge River, 153, 156.
Umsajbetsi River, 148.
Umsawe River, 177.
Umshane River, 150.
Umswezi River, 266, 270, 279.
Umtala, 283, 294.
Umzingwani River, 145, 148, 310.
Ushant, off, 5.

VAAL RIVER, 76.
Vehicle drawn by bullocks in Madeira, 9.
Vehicles for African Expedition, 120.
Veldt, the, from Kimberley to Vryburg, 50-51; round Pretoria, 83-97; round Pietersburg, 101-2; lost in the, 138-141, 150-2, 157, 251-3; bush veldt, 179; fire, 187-8, 204, 212, 249; between Fort Salisbury and Hartley Hill, 258.
Viandt, a Boer named, 183-4.
Victoria Falls of the Zambesi, 50.
Vigilance Committee at Fort Salisbury, 287-9.

INDEX.

Vryburg, 49-52, 121, 312, 326-7.

WAGES at De Beers Mine, 39 ; at Robinson Mine, 68.
Walden (servant), 166-7.
Wanetse River, 177-9, 302, 308-9.
Warm Baths near Pretoria, 98.
War Office, the, 30, 31.
Warren, Sir Charles, and Bechuanaland, 124.
" Warrigal" mine, the, 297.
Water on the plain, 196-8.
Water rates in Transvaal, 62.
Wegdraai, 132.
Wellington, town of, Cape Colony, 34.
Wells, 202, the Boers and, 102.
Wilderness, lost in the, 138-141, 252-3.
Wildebeest, 75, 76, 199, 218.
Wild pig, 147, 218, 232.
Williams, Mr. Gardner, 40, 132-4.
Williams, Captain G., 4, 97, 148, 153-9, 166, 169-173, 176, 185, 211, 234, 238-241, 253, 265, 280.
Willoughby, Sir John, 206, 209, 211, 218-233, 264-5, 287.
Wines of the Paarl, 34.
Winslow, Mr., 130.

Winton, Sir Francis de, 93.
Witwatersrand, gold-field of, 327-9.
Worcester, town of, Cape Colony, 34.
Workings, old mine, in Mashonaland, 240, 303.
Wynberg, near Cape Town, 21.

YELLOW JACKET MINE, the, 242-3, 296, 307.

ZAMBESIA, 22.
Zambesi River, 209, 213, 320.
Zambilli, Queen, 28.
Zimboe River, 258, 263.
Zoological Gardens at Lisbon, 6-7.
Zoology : see *antelopes, baboons, blesbok, bucks, chimpanzees, crocodiles, elands, elephants, gazelles, giraffes, hartebeest, hippopotami, hyænas, jackals, koodoos, leopards, lions, monkeys, quaggas, rhinoceros, springbok, wildbeest, wild pigs.*
Zoutspanburg, mining district of, 99, 100, 159.
Zululand, the government of, 26.
Zumbo, 209, 232.

www.ingramcontent.com/pod-product-compliance
Lightning Source LLC
Chambersburg PA
CBHW051742300426

44115CB00007B/663